Strategic Reflections
Operation *Iraqi Freedom*
July 2004–February 2007

By George W. Casey, Jr.
General, U.S. Army
Retired

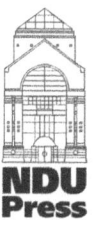

National Defense University Press
Washington, D.C.
October 2012

Opinions, conclusions, and recommendations expressed or implied within are solely those of the contributor and do not necessarily represent the views of the Defense Department or any other agency of the Federal Government. Cleared for public release; distribution unlimited.

Cover image: U.S. Army Sergeant Nathaniel Patterson, of the 320th Battalion, 3rd Brigade Combat Team, 101st Airborne Division, Mahmudiyah, Iraq (U.S. Army/Richard Del Vecchio).

*To the men and women from 38 countries, represented by the
Soldier on the cover, and the families that supported them, who
gave 27 million Iraqis the opportunity
for a better life.*

Contents

List of Illustrations x
Acknowledgments xi
Introduction 1

Chapters

**1 PREPARING FOR THE MISSION
    (MAY–JUNE 2004)** 5
   Framing the Mission 6
   Building a Key Relationship 10
   Consultations and Direction. 12
   Direct Insights 14
   Confirmation 16

**2 ESTABLISHING THE MISSION AND PREPARING
    FOR THE FIRST ELECTIONS
    (JULY 2004–JANUARY 2005)** 19
   Organizing for Success 33
   Military Operations. 40
   Election Planning and Preparation. 42
   Looking Beyond the Elections 44
   The First Elections: January 30, 2005. 49

**3 COMPLETING THE UN TIMELINE AND SETTING
    CONDITIONS FOR A NEW IRAQ
    (JANUARY–DECEMBER 2005)** 51
   Adapting in Transition. 55
   Completing the UN Timeline. 66

CONTENTS

4 GOVERNMENT TRANSITION AND THE RISE OF SECTARIAN VIOLENCE (JANUARY–JUNE 2006) 81

Government Transition 84

The Samarra Bombing and Its Aftermath 89

Building New Partnerships 98

Camp David and June D.C. Consultations 104

5 THE TOUGHEST DAYS (JULY 2006–FEBRUARY 2007) 109

Adjusting the Plan 110

Civil-Military Relations 120

Military Operations in Baghdad 126

Washington Policy Review 135

6 INSIGHTS FOR LEADERS 153

Developing Vision and Strategy 154

Creating Unity of Effort 156

Continuous Assessment and Adaptation 159

Influencing Organizational Culture 162

Civil-Military Interaction 165

Political-Military Integration 170

Momentum and Transitions 172

Sustaining Yourself 175

Appendices

    1. Abbreviations....................................179

    2. The Coalition, July 2004............................183

    3. Leadership in Iraq, 2004–2007 184

    4. Iraqi Security Forces Progression, 2004–2007...........188

    5. Operation *Iraqi Freedom* Chronology, 2004–2007 192

Notes..203

About the Author 207

Index..209

Illustrations

Figures

- 2-1. Multi-National Force–Iraq, July 2004 23
- 2-2. View of the Threat, Summer 2004....................27
- 2-3. Initial Campaign Framework 31
- 2-4. MNF-I Staff Organization 35
- 2-5. Counterinsurgency Practices........................45
- 3-1. Army Partnership Alignment 59
- 3-2. Commanding General's FARs, February 2005 63
- 4-1. Post-Samarra Security Environment, March 2006 93
- 4-2. Strategy for Post-Samarra Security Environment.......95
- 4-3. Campaign Framework, April 2006–December 2009........................ 97
- 5-1. Framework for Integrating Political-Military Efforts, Fall 2006 127

Graphs

- A4-1. Iraqi Security Forces Growth 189
- A4-2. Army Transition Readiness Assessment Growth.....190
- A4-3. Special Police Transition Readiness Assessment Growth 191

ACKNOWLEDGMENTS

This book would not have been possible without the efforts of Sandy Cochran and Kelly Howard, who have been involved in this effort since Iraq and who assisted me in researching, writing, and fact-checking. I am also appreciative of the comments from many who read the paper, and of the support of the National Defense University Press, particulary Frank G. Hoffman, Jeffrey D. Smotherman, and Tara J. Parekh. Finally, my wife, Sheila, was my strongest supporter throughout my career and particularly during my time in Iraq. She is also a tireless advocate for the families of the men and women of our Armed Forces.

INTRODUCTION

Operations *Enduring Freedom* and *Iraqi Freedom* were the first major wars of the 21st century. They will not be the last. They have significantly impacted how our government and military think about prosecuting wars. They will have a generational impact on the U.S. military, as its future leaders, particularly those in the ground forces, will for decades be men and women who served in Iraq and Afghanistan.* I believe it is imperative that leaders at all levels, both military and civilian, share their experiences to ensure that we, as a military and as a country, gain appropriate insights for the future.

As the Army chief of staff, I encouraged leaders at the war colleges, staff colleges, and advanced courses to write about what they did in Iraq and Afghanistan so that others could be better prepared when they faced similar challenges. This book is my effort to follow my own advice. I believe that we have not seen the last of the challenges and demands that I faced during my 32 months of combined command in Iraq, and I offer these insights so that future leaders can be better prepared.

During my command tenure, the Multi-National Force–Iraq (MNF-I) mission transitioned from one of occupation to one of partnership with three Iraqi governments. We used this period to lay the foundation for and begin the transition to a self-reliant Iraqi government, our mandate from the United Nations (UN). This book primarily addresses the events, decisions, and actions of MNF-I as I perceived them at the time. I have occasionally offered retrospective

---

* The last U.S. Army general with Vietnam experience retired in 2011.

insights that come from experience and hindsight, but I felt it is more important to focus on what I thought and what I did then to provide the best insight into the challenges I faced and how I dealt with them. It reflects my insights as a commander in one theater of a broader war.

In preparing this book, I relied on the historical records that I kept from Iraq—personal notes, briefings, plans, assessments, meeting notes—that, while decidedly incomplete, greatly sharpened my personal recollections and offered insights into how we viewed the situation over time and what we conveyed to our political leaders. They are available for research at the National Defense University.

As in any major endeavor, personalities mattered. Throughout my entire command tenure, I interacted with an extremely professional group of civilian and military leaders. My Commander in Chief was President George W. Bush, who was served by two Secretaries of State, Colin Powell and Condoleezza Rice, and two National Security Advisors, Dr. Rice and Stephen Hadley, whom I interacted with regularly. Strategic oversight came from two Secretaries of Defense, Donald Rumsfeld for 30 months, and Robert Gates for my final 2 months. I worked closely with two Chairmen of the Joint Chiefs of Staff, General Richard Myers, USAF, for 17 months and General Peter Pace, USMC, for 15 months. My immediate military commander for my entire tenure was Commander of U.S. Central Command General John Abizaid, USA. I was privileged to work side-by-side with two U.S. Ambassadors, John Negroponte and Zalmay Khalilzad, for 8 months and 22 months, respectively, and one interim Ambassador, Jim Jeffrey, who later returned as our Ambassador to Iraq. I worked with the leaders of three different Iraqi governments, led by Prime Ministers Ayad Allawi, Ibrahim al-Jafari, and Nuri al-Maliki. I had the benefit of working with very talented military subordinates: four Multi-National Corps–Iraq (MNC-I) commanders—then–Lieutenant Generals Thomas Metz, John Vines, Peter Chiarelli, and Raymond Odierno—and two Multi-National Security Transition Command–Iraq (MNSTC-I) commanders, then–Lieutenant Generals David Petraeus and Martin Dempsey. I also worked closely with Lieutenant General Stan McChrystal, who led our efforts against al Qaeda in Iraq and Afghanistan.

Over 65 years ago in his initial dispatch following Operation *Torch*, General Dwight Eisenhower wrote that "the accomplishments of this campaign are sufficiently evident to make comment unnecessary. Our mistakes, some of which were serious, may be less apparent at this moment, and, in the interest of future operations, they should be subject to dispassionate analysis."

Although the accomplishments of this campaign may not currently be seen with sufficient clarity to make comment unnecessary, I submit this book in the same spirit. The challenges that I faced during my command hold valuable lessons for future military and civilian leaders as we enter our second decade of war. It is my hope that this book will provide insights that allow future leaders to better prepare themselves for the challenges they will surely face in this era of persistent conflict.

I purposely focused this book on my actions and the actions of my headquarters and, as a result, have only touched briefly on the significant stories of the advances that were made during this time in training indigenous security forces, targeting high-value individuals, detainee operations, reconstruction, and dealing with improvised explosive devices. The men and women of the MNF-I, the Intelligence

Community, and the Department of State who served in Iraq during this time rewrote the books in these and other areas and postured us for success in Iraq and in future conflicts.

Operation *Iraqi Freedom* is part of a larger story—that of the United States as a nation adapting to the security challenges thrust on us by the al Qaeda attacks of September 11, 2001, and that of a military transforming in the midst of war. As this book illustrates, the forces involved, both military and civilian, adapted under fire and in the face of the uncertainty and complexity of Iraq to accomplish our national objectives and provide 27 million Iraqis the opportunity for a better life. It is a historic accomplishment, and one of which all Americans can be justifiably proud.

—George W. Casey, Jr.
General, U.S. Army, Retired
May 2012

President George W. Bush addresses U.S. Army War College on Iraq, May 24, 2004
White House (Eric Draper)

1. PREPARING FOR THE MISSION (MAY–JUNE 2004)

I did not go to work on May 17, 2004, thinking I would be the commander of Multi-National Force–Iraq in 45 days. I knew that the Secretary of Defense was looking to increase the new MNF-I headquarters from a three-star to a four-star command to handle the wide range of strategic issues that a corps's headquarters is not equipped to deal with. But as the Army vice chief of staff, I was decisively involved in the multiyear reorganization of the Army. I had been asked by my boss, Army Chief of Staff Pete Schoomaker, to see that through.

Three days later, Pete told me that I had been selected by President George W. Bush to lead coalition forces in Iraq. With Pete's support, I immediately shifted gears and laid out a plan to take command. My plan involved reading to update my thinking on counterinsurgency operations and the region; meeting with key figures in the executive branch to understand what was expected of me; meeting with leaders from other government agencies to understand how they planned to contribute; visiting the intelligence agencies to develop a better understanding of the intelligence picture (at least as it was viewed from Washington); meeting with knowledgeable experts outside of government to better understand the context for the current situation; meeting with financial and contracting experts to understand the mechanisms required to get the reconstruction effort moving; and meeting with numerous Members of Congress to get their views and to prepare for my confirmation hearing. This process proved essential in framing my understanding of the mission and greatly facilitated the rapid production of our campaign plan once I assumed command.

Framing the Mission

Years of experience at the strategic level had taught me that the higher up you go, the less guidance you receive. This mission proved no exception. I found there were three key documents that were most useful in framing the mission for Iraq: the National Security Presidential Directive (NSPD) of May 11, 2004, the President's May 24 speech at the Army War College, and United Nations Security Council Resolution (UNSCR) 1546, with attached letters from newly appointed Prime Minister of Iraq Ayad Allawi and Secretary of State Colin Powell.

The NSPD established the organization for U.S. Government operations in Iraq after the termination of the Coalition Provisional Authority, which was to occur no later than June 30, 2004, and the reestablishment of "normal" diplomatic relations with a sovereign Iraq. It made the Ambassador responsible for the "direction, coordination and supervision of all United States Government employees, polices and activities in country" except for the "U.S. efforts with respect to security and military operations in Iraq," which were the responsibility of the commander of U.S. Central Command (USCENTCOM), the combatant commander to whom I would report. It directed the "closest cooperation and mutual support" between them.[1]

The NSPD also designated the Secretary of State as responsible for the "continuous supervision and general direction of all assistance in Iraq" and directed the USCENTCOM commander to lead the efforts to organize, train, and equip the Iraqi security forces (ISF) "with the policy guidance of the Chief of Mission." It established two new organizations: one under the Secretary of State (the Iraq Reconstruction Management Office) to guide the development effort, and one under the Secretary of Defense (the Project and Contracting Office) to provide contracting and project management support to the reconstruction and assistance missions. Finally, the NSPD recognized that assisting Iraq through the transition to democracy would take "the full commitment of all agencies" of the United States, and enjoined the heads of all agencies to support the mission.[2]

Clear division of labor and lines of command are critical to the effective prosecution of any mission, and this NSPD endeavored to provide that. In retrospect, while the division of labor was clear, the NSPD did not create the unity of command necessary for the effective integration of civil-military efforts in successful counterinsurgency operations. The Ambassador and I would have to create the unity of effort required for success. This would prove a constant struggle as the two supporting bureaucracies—State and Defense—often had differing views. Things would get more complex as we increasingly brought the new Iraqi government into the effort. The political and economic effects, so necessary to sustaining our military success, would be outside of my direct control.

Shortly after the NSPD was issued, President Bush outlined our mission in a speech at the U.S. Army War College. He stated that our goal was "to see the Iraqi people in charge of Iraq for the first time in generations," and that our job in Iraq was not only to defeat the enemy, but also "to give strength to a friend—a free, representative government that serves its people and fights on their behalf." He laid out five steps to accomplish our goal:

- hand over authority to a sovereign Iraqi government
- help establish stability and security
- continue rebuilding Iraq's infrastructure
- encourage more international support
- hold free, national elections [that will bring forward new leaders empowered by the Iraqi people].[3]

President Bush noted that national elections were the most important of the five steps and that, because of recent violence in Fallujah and the South, we would maintain our troop level at 138,000 "as long as necessary." He stated that the United States would do "all that is necessary—by measured force or overwhelming force—to achieve a stable Iraq." These were comforting words to a prospective commander. Finally, he talked about accelerating our program for training Iraqi security forces with an eventual goal of an Iraqi army of 27 battalions and an overall ISF number (to include police and border guards) of 260,000,[4] making it clear that this would be a major part of my mission. In all, this seemed like clear direction, and I used the speech as the basis for my planning.

Perhaps the most important document in framing the mission was UNSCR 1546. It provided the chapter VII mandate from the United Nations: "... the Multinational Force shall have the authority to take all necessary measures to contribute to the maintenance of security and stability in Iraq."[5] The accompanying letters relayed the public consent of the new Iraqi government to accept MNF-I and the political transition laid out in the UNSCR. This public acceptance would be essential to me when it came to working with the Iraqi government. It also established a timeline for the political transition:

- forming the sovereign Interim Iraqi Government (IIG) that would assume governing responsibility and authority by June 30, 2004
- convening a national conference reflecting the diversity of Iraqi society
- holding direct democratic elections by December 31, 2004, if possible and in no case later than January 31, 2005, for a Transitional National Assembly, which would have responsibility for forming an Iraqi Transitional Government (ITG) and drafting a permanent constitution for Iraq leading to a constitutionally elected government by December 31, 2005.[6]

This gave the Iraqis and coalition forces a political timeline for the next 18 months, which we saw as a good, if not necessary, driver to force consensus on what we knew would be tough issues. What we did not anticipate was the debilitating effect that three governmental transitions would have on our efforts to increase the capacity of Iraqi institutions.

Finally, the UNSCR and its supporting letters clearly stated *my* responsibility to establish a "security partnership" with the soon-to-be sovereign government of Iraq and to assist in building the capability of the Iraqi security forces and institutions that, the UNSCR envisioned, would "progressively play a greater role and ultimately assume full responsibility for the maintenance of security and stability in Iraq."[7] The UNSCR gave me a direct role with the sovereign government of Iraq to coordinate this security partnership, a role normally reserved for the Ambassador. I did not realize at the time how difficult and all-consuming this particular task would become.

Building a Key Relationship

I recognized from the outset that a close, cooperative relationship between John Negroponte, the newly appointed Ambassador to Iraq, and me would be absolutely essential—an instinct that he shared. We worked hard from the beginning to ensure that we entered Iraq with a common view of the situation and how we needed to address it. One of the most important agreements we made took place at our first meeting. There we discussed the fact that any counterinsurgency effort required political and military integration for success, and we agreed upon a concept to create unity of effort between the Embassy and MNF-I—*One Team/One Mission*. We agreed that we would develop a common statement of our mission and then guide the Embassy and MNF-I teams to work together to accomplish it. This understanding would prove vital to our success.

One of the toughest challenges for strategic leaders is to clearly articulate to their subordinates what it is they want them to accomplish. Before we left, the Ambassador and I worked to develop a clear view of what we wanted to accomplish in Iraq—understanding that we would take a period of time after we arrived to calibrate our views with realities on the ground. We also discussed the NSPD, the President's speech, and the UNSCR and how they would help us frame what we needed to do. We recognized that the return of sovereignty to the Iraqis presented both challenges and opportunities, and we wrestled with how to use the transition to create momentum for the mission. To do this, we felt we needed to work on enhancing the legitimacy of the IIG to move Iraqis away from the perception of the coalition as an occupying force. We also realized that the transition from the Coalition Provisional Authority to the Embassy and the One Team/One Mission concept would require some significant organizational changes to enable our success, and we began planning how to accomplish them.

In the end, we went into Iraq thinking that our mission was to facilitate the establishment of a representative Iraqi government that respected the human rights of all Iraqis, and that had sufficient security forces to maintain domestic order and deny Iraq as a safe haven for terrorists. To achieve that objective, we knew that we would have to build the national and international team to accomplish our mission, develop an integrated effort to defeat the insurgency, and work to build the legitimacy of the IIG and ISF.

These discussions with the Ambassador were extremely helpful in establishing a common view of the mission and the challenges we would face together. It was, we both realized, just the beginning of a long journey, but we were at least starting in the same place.

Consultations and Direction

As part of my preparations, I solicited views on Iraq from various experts from inside and outside the government: the Department of Defense, Joint Chiefs of Staff, Department of State, National Security Council, and Intelligence Community. The Institute for National Strategic Studies at the National Defense University and my alma mater, Georgetown University, both hosted special sessions for me that were very helpful.

My consultations with these organizations surfaced a wide range of concerns and questions. Some experts questioned how to obtain and sustain unity of effort between the Embassy and military, while others wondered about the challenges and implications of sovereignty. There were concerns about the newly constituted ISF (mission, force levels, equipment requirements, and timelines for development), and the impacts of disbanding the Iraqi army, stringent de-Ba'athification policies, and Abu Ghraib. There was also real uncertainty about the nature of the threat. While most agreed that we were dealing with an insurgency, there was much debate about the composition of the insurgency. Lastly, from these consultations I gained a sense that people thought that Iraq would be an 18-month mission: we would complete the UNSCR political timeline while growing the ISF and turn the country over to the Iraqis when that was done. In all, I found that having access to a wide variety of views and insights better helped me sharpen my thinking about the mission.

During that month, I had several office calls with Secretary Rumsfeld and Chairman of the Joint Chiefs of Staff General Dick Myers to get direction. I had reviewed the Secretary's April 27, 2004, guidance to USCENTCOM that planners should maximize the use of ISF, international forces, and contractors before resorting to U.S. forces.[8] He also sent me a copy of the memorandum he prepared for the President in early June, entitled "Some Thoughts on Iraq and How to Think about It," that sought perspective from history with respect to what he termed "a rough period of months." He emphasized that "there is no way this struggle can be lost on the ground in Iraq. It can only be lost if people come to the conclusion that it cannot be done."[9] Those were prescient words.

During these office calls, the Secretary emphasized two concerns. The first was about the "can-do" attitude of the American soldier. The Secretary was worried that, in our zeal to accomplish the mission, we would try to do everything ourselves and not allow the Iraqis to gain the experience they would need to ultimately take charge. He felt that this would only extend our time there, and he encouraged me to take this attitude into consideration in my planning. I understood what he meant, having seen this attitude in our soldiers in Bosnia, and even getting captured by it myself during my time there. We were going to have to find the right balance between the drive needed to accomplish things in a tough environment and doing everything ourselves if we wanted the Iraqis to take charge anytime soon. This would be easier said than done. Secretary Rumsfeld and General Myers were also concerned about the status of the ISF, and they asked me to develop an immediate assessment and long-term plan for ISF development as a matter of priority. We agreed that I would report back with an assessment of the situation and recommendations within my first 30 days on the ground.

Direct Insights

In mid-June I was granted permission from the Senate Armed Services Committee to accompany Deputy Secretary of Defense Paul Wolfowitz on a 5-day trip to Iraq.* The purpose of the trip was to gain a better understanding of the issues surrounding the transition of sovereignty to the IIG,† which was scheduled to take place by the end of the month. The trip would enable me to meet with the Iraqi and coalition military leaders whom I would be working with to gain important on-the-ground insights. I focused on gaining an understanding of how the new Iraqi leaders viewed the threat, their current security challenges, their security forces, and the consultative mechanisms called for in the UNSCR to ensure coordination between the coalition and the Iraqi government. Not surprisingly, the insights I gained from this visit played a prominent role in preparing me to take command of the mission. Here are a few of my key takeaways from the trip:

Threat and Security Challenges. Prime Minister Allawi viewed radical Islamists and ex-regime loyalists, who were increasingly siding with the radicals, as Iraq's primary threat. He thought both groups were getting support from regional powers, primarily Syria and Iran, and taking advantage of Iraq's porous borders to undermine the

---

* Permission was necessary to avoid "presumption of confirmation." We can do better preparing senior leaders for key wartime jobs. For example, I would have welcomed the chance to study Arabic for several months, something I could not do without "presuming confirmation."

† The Iraqi Interim Government was appointed by the Coalition Provisional Authority just prior to sovereignty being passed.

political process under way. He stated that things would get worse before they got better and that establishing a functional democracy in Iraq would take a long time. He also said that his priority was to establish security across Iraq. It was clear that the April uprisings by Sadrist militia and the failed efforts to establish a government security force in Fallujah weighed heavily on him and the new government. Muqtada al-Sadr had established a safe haven in Najaf, and terrorists and insurgents had established a safe haven in Fallujah. Coalition and Iraqi forces loyal to the central government could not go into either area. The Iraqis saw them as separate problems with Fallujah being the more serious of the two. They also saw them as longer term problems and did not expect them to be resolved before sovereignty was established. I would inherit them.

Vision for Iraqi Security Forces. The prime minister and his security ministers believed there were insufficient ISF to deal with the threat, and those that did exist were underequipped. They saw this situation as unacceptable, and rightly so. On the army side, they looked down on the recently formed Iraq Civil Defense Corps (coalition-armed local security forces), and the prime minister and his ministers felt that they needed armored forces—at least five divisions—that could rapidly deploy around the country. They also wanted an aerial capability to assist in the counterinsurgency fight. On the police side, they recognized that, given the threat, the police would need the support of the army for some time, and that current training needed to be enhanced to allow the police to survive in a counterinsurgency environment. They wanted to create strong border and counterterror forces. They also wanted our help to unify the security effort (coalition, army, police), to develop an appropriate chain of command for the army and police, to build a "rapid deployment force," so the central government could respond anywhere in the country, and to develop a strict vetting process for key leaders. Finally, they wanted *Iraqi* forces, not "photocopies of the U.S. or UK forces."

Consultative Mechanisms. The letters from Prime Minister Allawi and Secretary of State Powell attached to UNSCR 1546 called for the establishment of consultative mechanisms to facilitate coordination between the coalition and the sovereign government of Iraq. We agreed that the Ministerial Committee for National Security would be the core forum for working strategic security issues and that the Strategic Action Committee would be the forum to prepare issues for its consideration. We began discussions on developing a policy for "sensitive offensive operations"—operations that could cause political problems for the government—and establishing formal and informal coordination mechanisms at the national, provincial, and local levels. We also agreed to establish a joint command center as quickly as possible. Establishing these agreements in advance would be critical to progress in the months ahead.

Confirmation

The conclusion of these busy weeks came with my confirmation hearing before the Senate Armed Services Committee on June 24 shortly after my return from Iraq. During the several days prior to the hearings, I visited key members of the committee to get their views and insights on the mission, and submitted my "advance questions" to the committee for the hearing.

I felt that my preparations to take command had set me up well for the hearing as the Senators echoed many of the concerns I had been hearing in the past few weeks. Members of the committee asked me how I planned to ensure unity of effort with the Embassy and to establish a good relationship with the Ambassador, how I planned to establish an effective relationship between MNF-I and the government of Iraq, about the status of the ISF and how I planned to develop them, and how I viewed my relationship with General Abizaid. I was also pressed by a number of Senators on whether I felt I had enough troops to accomplish the mission. I pledged several times to ask for more troops if I felt they were necessary, but I reiterated, as I did with many of the questions, that I had only been on the ground in Iraq for 3 days, and I would make a thorough assessment once I took command. I also agreed to consult with them frequently.

In response to a question posed by the committee concerning the major challenges I would face as the MNF-I commander, I listed the following:

- implementing an effective transition from occupation to partnership with the IIG
- defeating anti-Iraqi and anticoalition forces alongside the IIG and ISF
- assisting the IIG in efficiently rebuilding the ISF
- with the ISF, providing a secure environment to permit elections in December 2004 or January 2005.[10]

This represented an accurate view of what I thought my main challenges would be as I prepared to depart for Iraq.

My preparation time for command had been brief but intense. I left for Iraq with a good idea of what was expected of me and what needed to be done. From the United Nations, I had received an 18-month political timeline to execute. President Bush provided the goal of seeing "the Iraqi people in charge of Iraq for the first time in generations" along with a five-step framework to accomplish this.[11] Secretary Rumsfeld and General Myers had asked me to report back with an assessment and way ahead for not only the ISF but also for the mission in Iraq as a whole.

Going in, I believed that the U.S. objective was to facilitate the establishment of a representative Iraqi government that respected the human rights of all Iraqis and had sufficient security forces to maintain domestic order and deny Iraq as a safe haven for terrorists. I knew that I needed to quickly make an on-the-ground assessment, develop a strategy and a campaign plan to achieve our objective, and then work with Ambassador Negroponte to build our team and organize the mission for success—all while working to build a strong partnership with the newly sovereign Iraqi government. We would have plenty to do. While I knew the mission in Iraq would not be easy, I was just starting to understand its complexity.

General John P. Abizaid, Commander, U.S. Central Command, General Casey after he assumed command of Multi-National Force–Iraq, July 1, 2004, and outgoing commander Lieutenant General Ricardo Sanchez, seated behind General Abizaid

2. ESTABLISHING THE MISSION AND PREPARING FOR THE FIRST ELECTIONS (JULY 2004–JANUARY 2005)

On June 28, as I was about to board a plane to the Middle East having been confirmed by the Senate 2 days prior, Ambassador Negroponte called to tell me that sovereignty had been passed to the Iraqis earlier that day. While he intended to head into Iraq later that day, I had planned stops at USCENTCOM forward headquarters in Qatar and at my supporting Army headquarters in Kuwait, Third Army, en route to Iraq. Despite the early transfer of sovereignty, I decided to stick with my travel plan as insights from my higher and supporting headquarters in theater would be important in framing my understanding of the mission. My change of command remained scheduled for July 1.

I arrived in Iraq the night of June 29 and immediately began meeting with the key people in the mission: outgoing commander Lieutenant General (LTG) Ricardo Sanchez, Ambassador Negroponte, my United Kingdom (UK) deputy Lieutenant General John McColl, and new Multi-National Security Transition Command–Iraq Commander LTG David Petraeus. The following day, I conducted my first secure video teleconference with President Bush and his national security team in Washington. I told the President that I would give him my assessment of the overall situation and recommendations for the way ahead in 30 days and that my immediate priorities were to develop an integrated counterinsurgency strategy to defeat the insurgency, develop a plan for the formation of ISF, build the consultative and coordinating mechanisms with the IIG, and complete the transition of military support from the Coalition Provisional Authority to the Embassy. I assumed command of MNF-I the following day, July 1.

Following the change of command, I met with my immediate boss, General Abizaid, to receive his oral and written guidance for the mission. He had been in USCENTCOM for 18 months and was commander for the last year. I would be one of his two theater commanders (LTG Dave Barno was the commander in Afghanistan). John was a seasoned regional hand and a close friend whose insights I valued. His direction reflected his experience. He told me to focus on setting the conditions for the January elections while building loyal Iraqi security forces and institutions and respecting Iraqi sovereignty. He told me to let him know the adequacy of the rules of engagement and support from his headquarters, and informed me that I was authorized to communicate directly with the Chairman and Secretary of Defense on "matters relating to the operational and tactical direction of the force." He asked only to be kept informed in these instances. This would substantially increase our agility to prosecute tactical actions, and I resolved not to abuse this trust. Our session began an invaluable relationship that continued throughout my entire tenure.[1]

At that time, MNF-I consisted of around 162,000 coalition forces from 33 countries that had been organized into five Multi-National Division (MND) areas of operation and one Multi-National Brigade (MNB) area of operation in northwest Iraq (see figure 2-1). MND–South East was commanded by a UK two-star general, and MND–Center South was commanded by a Polish two-star general. These two divisions contained the preponderance of non-U.S. coalition forces. MND-Baghdad, MND–North Central, and MNF-West, the USMC sector, were commanded by U.S. two-star generals, and MNB–North West was commanded by a U.S. one-star general. While the U.S. units contained some multinational forces, they were predominantly U.S. organizations. These units reported directly to the Multi-National Corps–Iraq commander, a U.S. three-star general who was responsible for orchestrating the operational aspects of our mission. I visited each of these units and a good number of their subordinate brigades and battalions in the first 30 days after my arrival. Not surprisingly, the insights provided by subordinate commanders were invaluable in developing my assessment of the situation.

While the Ambassador and I crisscrossed the country meeting with Iraqi and coalition leaders to build our own assessment of the situation and refine our vision and strategy, we undertook two separate and parallel staff actions to help us frame our mission and plans.

First, I made a decision to continue with the campaign planning that had been initiated by my predecessor, LTG Sanchez, when the MNF-I headquarters was formed that May. The headquarters was established to provide a separate four-star theater headquarters to handle the strategic aspects of the mission and to deal with Washington, the Embassy, and the Iraqi government. This was a very necessary step, and over the course of the mission, it greatly facilitated the accomplishment of our national objectives. The MNF-I headquarters was established on May 15, 2004, with personnel authorizations for individual officers and noncommissioned officers from across the U.S. Services and coalition countries. These personnel were slow to arrive, and the headquarters was still forming when I arrived at the end of June.

To complement that nascent planning effort, the Ambassador and I felt we needed a way to bring our key subordinates and staffs together with a shared view of the threat, the nature of the conflict we were involved in, and our mission, so we decided to form a Red Team—a group of experienced senior people empowered to operate outside of normal staff processes to provide their insights and recommendations directly to the Ambassador and me. Our hope was that the Red Team would both provide us with alternative views that we could use to vet the MNF-I campaign plan and, just as importantly, form a basis for a joint mission statement, which the Ambassador and I would issue. This document would enable us to bring our respective organizations together around common objectives and operationalize the One Team/One Mission concept we had agreed to in Washington.

The Red Team was led by a senior Foreign Service officer with an Army two-star general as his deputy. Their task was to take an

ESTABLISHING THE MISSION

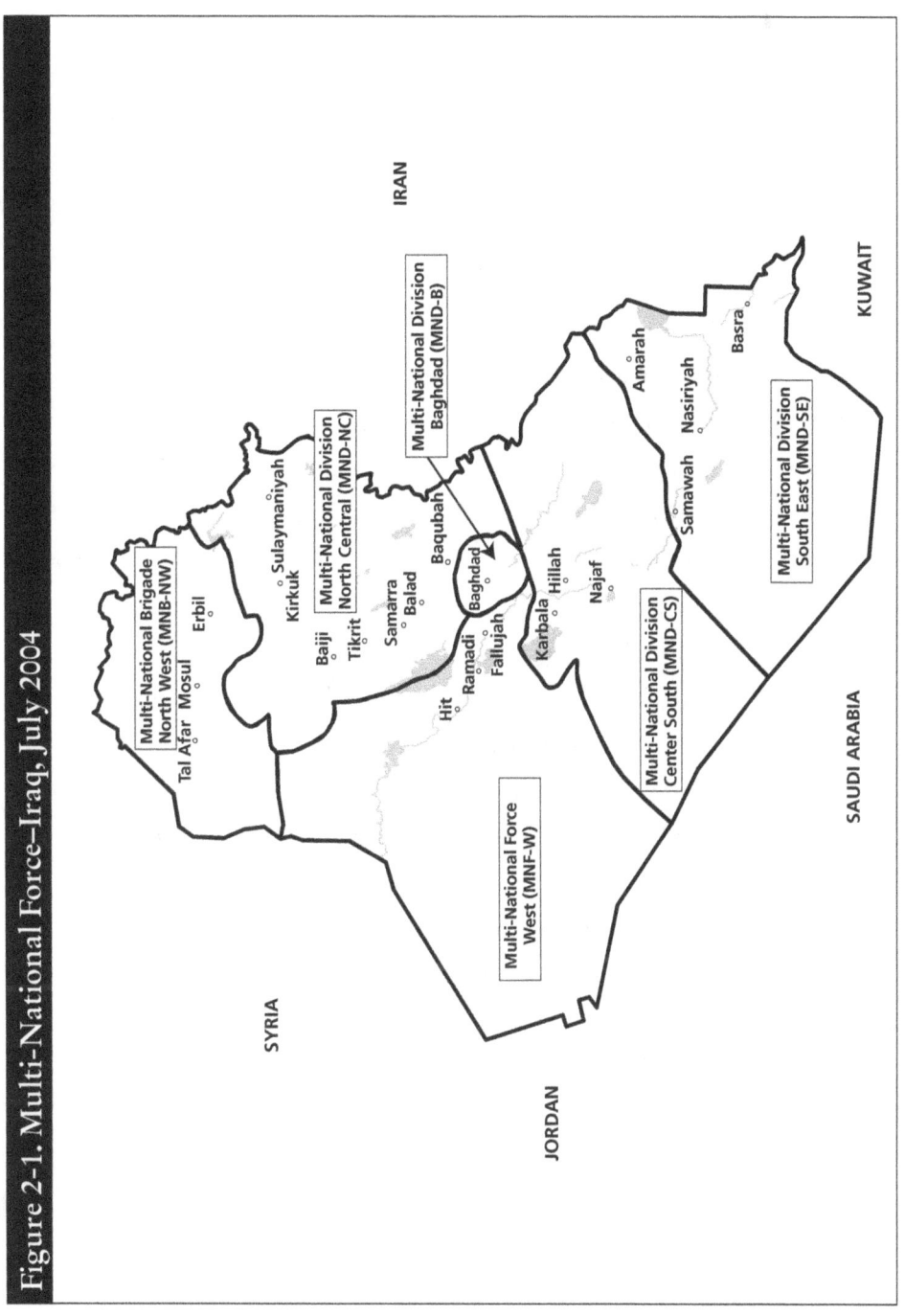

Figure 2-1. Multi-National Force–Iraq, July 2004

independent look at both the nature of the threat and the nature of the war, and to give us recommendations on how we should proceed. The team consisted of handpicked senior members of the Embassy and Central Intelligence Agency (CIA), the British embassy and Special Intelligence Service, and MNF-I. The Ambassador and I gave them 30 days to do their work, with the intent of bringing it together with the ongoing MNF-I campaign planning effort. We planned to issue the joint mission statement and campaign plan by early August. I felt very strongly that it was my responsibility to ensure that every member of the coalition clearly understood what it was that we were trying to accomplish in Iraq so each one could contribute to our success. These two documents would go far in helping me do that.

In late July, after several productive sessions with them, the Red Team reported back to the Ambassador and me. They concluded that we were fighting an insurgency and that it was "stronger than it was nine months ago and could deny the IIG legitimacy over the next nine months." In their view, the insurgency was primarily led by well-funded Sunni Arab "rejectionists" who had lost power with the overthrow of Saddam Hussein and rejected the new order. The rejectionists centered around former regime elements, members of the former Ba'ath Party, and former Iraqi security and intelligence forces who had the wherewithal to challenge the formation of a democratic government in Iraq. The Red Team felt that there was "not a monolithic Ba'ath Party" controlling the insurgency, but a "loose system of leadership with no single leader," and that many of the key leaders and facilitators were based outside of Iraq, primarily in Syria. The insurgents shared a range of motivations from "the explicitly religious to Arab nationalists to Saddam loyalists."

They felt that foreign Islamic extremists (al Qaeda) were a "small if lethal problem in Iraq" (numbering fewer than 1,000) and that "Iran is hoping to win influence over Iraq's political and electoral process without having to provoke a Shia-based insurgency (for which it is preparing, nonetheless)."[2] Despite their different objectives, all insurgents shared a common goal—the failure of the coalition mission. We accepted this view of the threat and graphically portrayed it in our campaign plan as shown in figure 2-2. The staff dubbed this representation "The Wonder Bread Chart."

The Red Team also concluded that "although the IIG enjoys early popular support, it has a weak hold on the instruments of governance and has to manage a war-battered economy, a fragile and damaged infrastructure and the meddling of some neighboring states, especially Syria and Iran." They noted that we, and the members of the international community, needed to work to strengthen the legitimacy of the IIG in the eyes of the Iraqis in order to strengthen the government's hand in dealing with these challenges and defeating the insurgency. They felt that our political, economic, and security efforts were "hampered by the lack of a unifying strategy, inadequate intelligence, ineffective strategic communications and the embryonic nature of IIG counterparts."[3] We clearly had our work cut out for us.

Looking back, the Red Team was an effective vehicle to bring together senior political, military, and intelligence leadership to address the key issues affecting the mission and how to deal with them. We agreed on broad issues, such as the nature of the enemy (Sunni Arab rejectionists), the nature of the war (counterinsurgency), the nature of our relationship with the Iraqi government (partnership), and our mission in Iraq (to help the Iraqi people build a new Iraq).

While everyone did not agree on everything, we at least all knew where we stood, and we were close enough on the major issues to get moving. Our effort did not, to this point, include the Iraqis—a gap that we would close over time. I was so pleased with the results of the Red Team effort that I used it frequently throughout my tenure to shed light on difficult issues.

Working directly from the Red Team assessment, the Ambassador and I crafted a joint mission statement for our respective organizations and signed it on August 18. In this first critical document, we formally defined our objective: "To help the Iraqi people build a new Iraq, at peace with its neighbors, with a constitutional, representative government that respects human rights and possesses security forces sufficient to maintain domestic order, and deny Iraq as a safe haven for terrorists." We stated that the IIG shared this objective, but was "in the early stages of consolidating the aspects of national power," so we aimed "to bolster the IIG's legitimacy in perception and fact," acknowledging this would be a major challenge. We also conveyed our common view of the threat, noting that the gravest immediate threat to IIG legitimacy was an insurgency principally led by well-funded Sunni Arab rejectionists drawn from former regime elements. To deal with that threat, we laid out a series of tasks in three interrelated categories: political, security, and economic, and asserted that these tasks would be the *"focal point of integrated efforts mounted by everyone operating in Iraq under our authority* [emphasis added]."[4]

The joint mission statement was a good start, but it was not sufficient to guide coalition military efforts in a multiyear campaign, especially one in which national contingents rotated once or twice a

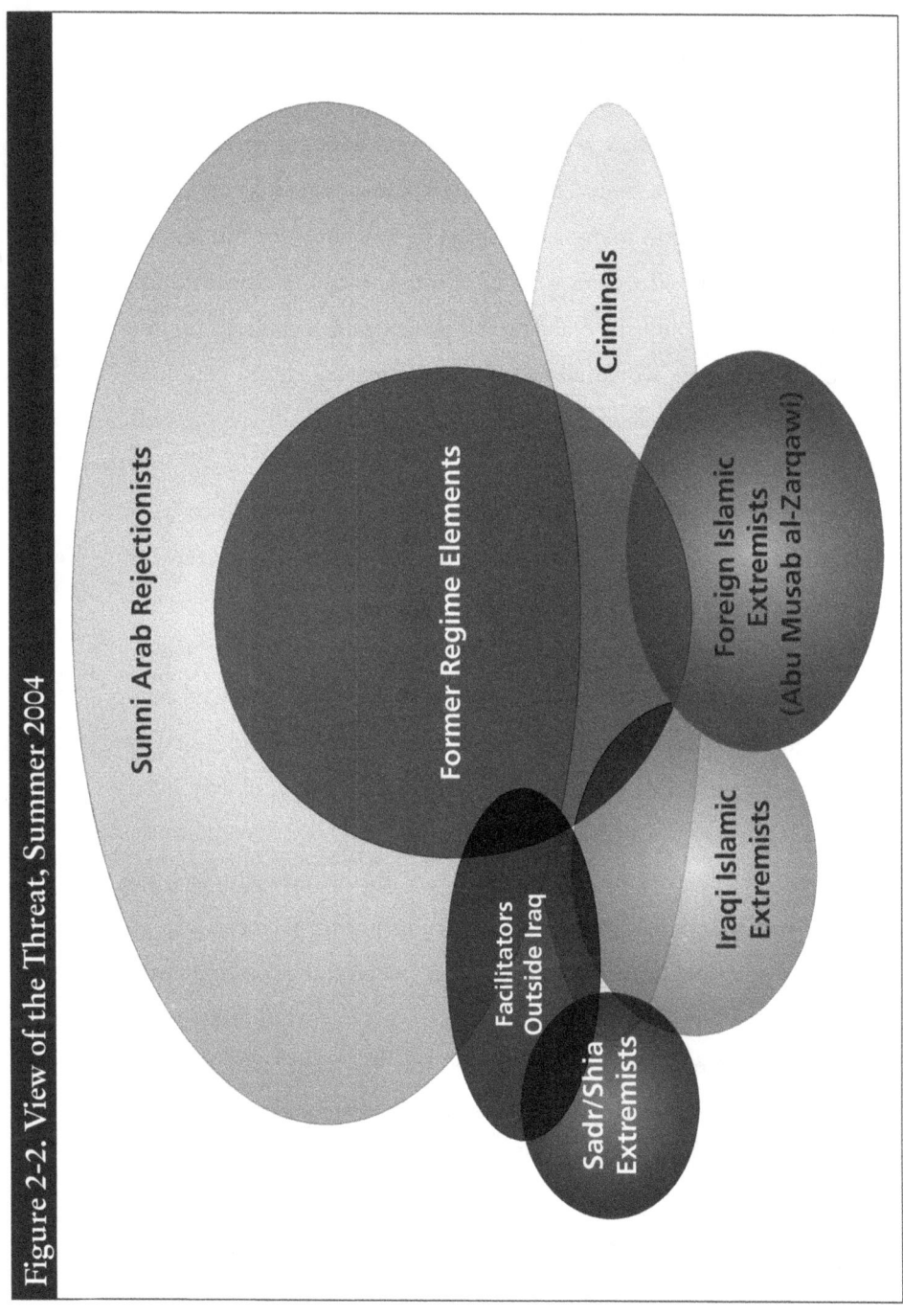

Figure 2-2. View of the Threat, Summer 2004

year. For that we needed a campaign plan. I intended to craft a written plan to clearly define our mission and how I saw the threat and risks and to articulate a strategy and organizational framework to accomplish the mission. This campaign plan would also provide operational direction to my subordinate commanders for the conduct of the military effort and would put in place an assessment mechanism to continuously evaluate our progress in accomplishing our objectives. This would allow us, again continuously, to reevaluate the conscious, and unconscious, assumptions that drove the plan, and to adapt it as necessary.

As I traveled throughout the country, my staff continued work on the campaign plan in parallel with the Red Team effort.* I met with the campaign planners several times a week to cross-level insights and discuss important issues. One of the key discussions we had was on the "center of gravity," an important element of any successful campaign. While we generally agreed that the strategic center of gravity was coalition public support, we differed on the center of gravity for the Iraq theater of operations. In counterinsurgency operations, the center of gravity is usually the people of the country in which the insurgency is being contested. Our discussion revolved around the issue of who could best "deliver" the Iraqi people—the coalition or the Iraqi government. I felt that, as our goal was a government seen as representative by the Iraqi people, the more we did to build the legitimacy of those governments in the eyes of the Iraqi people, the sooner we would achieve our goal. Others argued that we should focus more

---

* Although the joint mission statement was signed shortly after the MNF-I military campaign plan was released, good cross-staff coordination, and my personal oversight, ensured that its tenets were fully incorporated into the plan.

directly on the Iraqi people. In the end, we made the legitimacy of the Iraqi government the theater center of gravity. We spent a great deal of time debating this and several other key issues as we built the campaign plan, but it was time well spent. I found that the issues we were dealing with were so complex that I benefited from hearing different views when making critical judgments.

The campaign plan, issued August 5, 2004, laid out direction for the next 18 months. The plan put the Iraq mission in the context of our efforts up to that time (the Liberation and Occupation Phases of Operation *Iraqi Freedom*), and focused primarily on the next 18 months (the Partnership Phase), which entailed the completion of the UNSCR timeline and the formation of a constitutionally elected Iraqi government by 2006. The plan looked beyond January 2006, but only broadly, to the Iraqi Self-reliance Phase, where Iraqis would assume security responsibility. As we were still early in the mission, we purposely did not assign a timeline for this phase.

The mission statement from the campaign plan reflected the key elements of partnership with the IIG, counterinsurgency operations, training and equipping ISF, and completing the UNSCR 1546 by the end of 2005: "In partnership with the Iraqi Government, MNF-I conducts full spectrum counterinsurgency operations to isolate and neutralize former regime extremists and foreign terrorists and organizes, trains and equips Iraqi security forces in order to create a security environment that permits the completion of the UNSCR 1546 process on schedule."[5]

To accomplish this mission, we laid out a counterinsurgency strategy that sought to use the full spectrum of military and civilian tools to separate insurgents and extremists from the Iraqi people and defeat the insurgency while we restored Iraqi capacity to govern and secure the country. We knew executing this strategy would be very difficult in what amounted to a postwar failed state—although it would be awhile before we realized how difficult—so we laid out a framework in the campaign plan designed to integrate and synchronize all of the elements of Iraqi and coalition power to accomplish our objectives over time. We used four lines of operation representing the four major elements of power that we would bring to bear: security, governance, economic development, and communicating. Each of these lines was aligned with specific organizations designated to accomplish the specific effects shown in figure 2-3. We made a conscious effort to minimize what the U.S. Government sought to achieve as we developed these objectives, and believed that accomplishing these effects in an integrated fashion would lead us to the endstate.

Coordinating the integration of efforts would have been tough for any one organization, but our efforts were complicated by the fact that we had two organizations—the Embassy and MNF-I—that shared responsibility for success. MNF-I was responsible for security and the Embassy for governance and economic development. We shared responsibility for communicating. We also shared responsibility with the Iraqi government. Execution and coordination within and across the lines of operation were continuous challenges, underscoring why the One Team/One Mission concept was so important. Everyone had to deliver in a coordinated fashion if we were going to succeed.

On the security side, I told the MNC-I commander to conduct a counterinsurgency campaign to:

ESTABLISHING THE MISSION

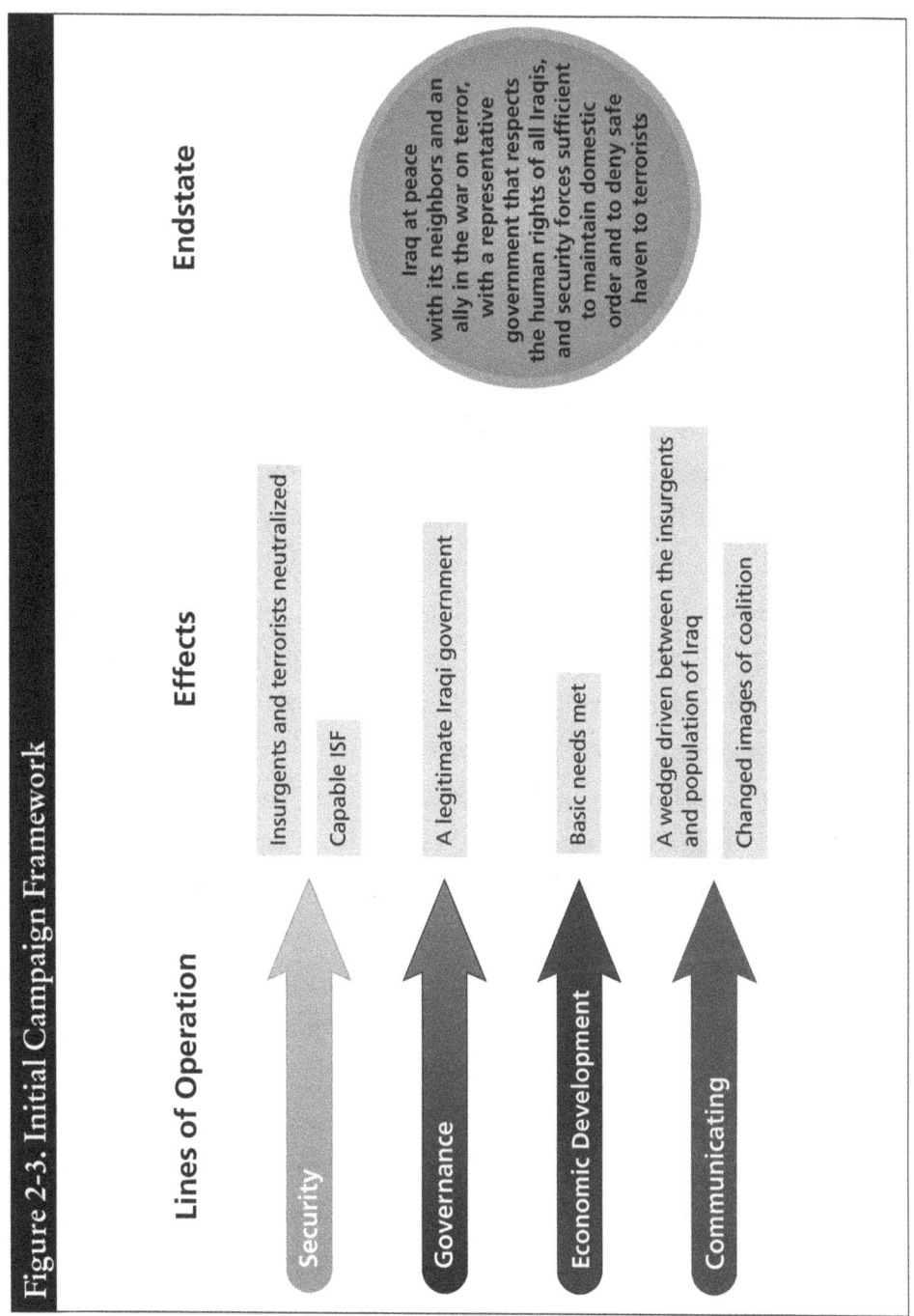

Figure 2-3. Initial Campaign Framework

- neutralize the insurgency in the Sunni Triangle
- secure Baghdad
- block the borders of Iraq to disrupt the flow of support to the insurgency
- assist in building the ISF
- sustain support for coalition force efforts in Shia and Kurdish areas.

With an eye toward getting an "Iraqi face" on elections in January 2005, I directed the MNC-I and MNSTC-I commanders to focus on getting the ISF to the point where they could plan and conduct security operations at the platoon/police station level with limited coalition support by January. I told them to prioritize these efforts in 15 key cities, in which almost half of the population of Iraq resided. I also gave them some broader ISF objectives for 2005, but our initial focus was successful elections in January.

The campaign plan also established the Commander's Assessment and Synchronization Board (CASB) to assess and manage the accomplishment of the plan. We recognized upfront that the campaign plan was a "living" document that would have to be adjusted as conditions changed and assumptions failed to materialize. The CASB was initially designed to be a monthly process, but we soon went to a bimonthly timeline to minimize redundancy and reduce staff time spent on preparations. I found that getting the assessment process to yield meaningful results—ones the Ambassador and I could act on—took a great deal of my personal effort. The tendency of a staff is to track the things that are most easily measured, not necessarily what is most critical. I finally found that if I forced the staff to answer the following three questions about each effect, I came closer to getting what I needed: What are we trying to accomplish? What will tell us if we are accomplishing it? How do we measure that? It took a year of trial and error before I was satisfied with the assessment process.

With the campaign plan complete, I briefed my staff and subordinate units and gave them 2 weeks to review and develop their supporting plans. I spent the latter part of August listening to back-briefs presented by my subordinates to ensure that they understood the plan and my intent. I was generally pleased with their work. I also shared the plan with the Ambassador and Embassy staff, and the Ambassador and I briefed President Bush and the National Security Council in mid-August. As part of this briefing, I highlighted to our leadership some "potential good and bad" things that could happen in the next 6 months that could affect the plan. I wanted to remind them that we were at war and that things would change. While I had been on the ground for only a month and had developed and issued a campaign plan, I knew we still had a very long way to go.

Organizing for Success

As the Ambassador and I looked at what we had to do, it was clear to us that the One Team/One Mission concept required some changes to both of our organizations to facilitate the integration of our efforts. For starters, we put our offices next to one another and met frequently over the course of the week. I looked at the configuration of the MNF-I headquarters—which was the standard J1–J9 organization that worked so well in conventional operations—and realized that it would not be suitable for executing the key functions of a counterinsurgency campaign plan where political, economic, and information effects needed to be generated and synchronized with the security effort, and vice versa. It was also clear that the MNF-I staff would have to work closely with the Embassy staff, and that this could not happen effectively if they operated from separate locations. I also had an internal MNF-I issue in that I needed to refocus my headquarters at the theater level and get them out of the corps's operational and tactical business, which they had been overseeing until the standup of the MNF-I headquarters in May.

After discussions with my staff and the Ambassador, and some help from U.S. Joint Forces Command, we designed a headquarters that could more easily carry out the nonstandard functions of the campaign and that would better facilitate the integration of the civil-military effort. To do this, we split the MNF-I headquarters between the Embassy in the Green Zone and Camp Victory in West Baghdad. To the Ambassador's credit, he accepted about 300 military personnel working permanently in his Embassy alongside his staff. These staff officers worked to integrate our security plans with the Embassy in the key areas of operations, planning, assessment, strategic communications, and reconstruction and economic development. We created three staff sections at the Embassy: Strategic Plans and Assessments, Political-Military-Economic Effects, and Strategic Operations, all under the oversight of my UK deputy and working directly with Embassy principals. At Camp Victory, we retained the key support functions (personnel, logistics, signal, intelligence) and detainee operations. I maintained offices in both locations, starting my day at Camp Victory, but spending the majority of my time working from the Embassy office or visiting units across the country. The wiring diagram in figure 2-4 lays out the organization that we established that summer.

ESTABLISHING THE MISSION

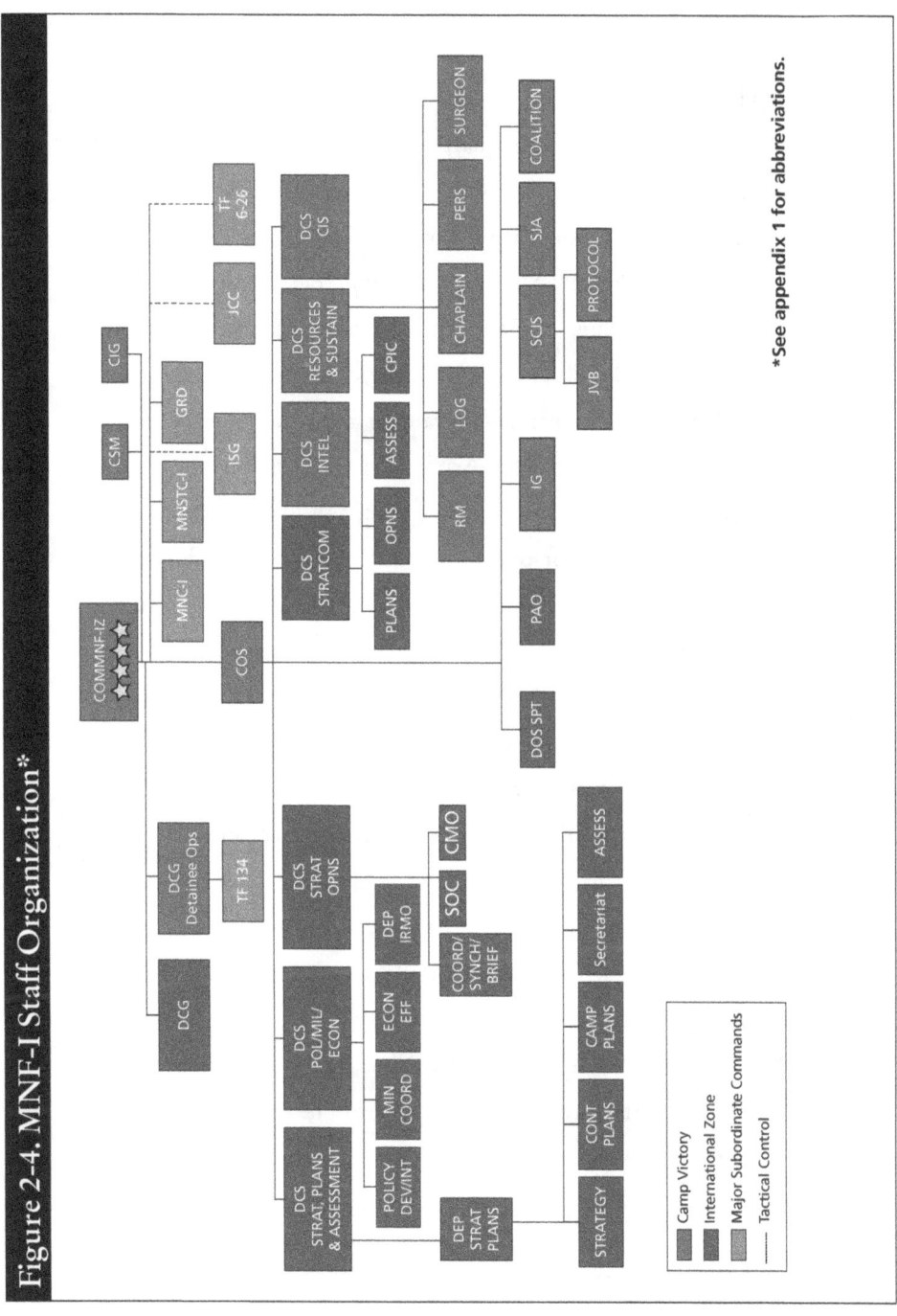

Figure 2-4. MNF-I Staff Organization*

*See appendix 1 for abbreviations.

STRATEGIC REFLECTIONS

No organization is perfect, but this organization served us well, with minor adjustments, during my tenure.

My key subordinate headquarters, their responsibilities, and locations were as follows: MNC-I, the corps's headquarters, responsible for the operational aspects of the mission; Task Force 134, responsible for detainee operations; and the Iraq Survey Group, responsible for searching for weapons of mass destruction until it disbanded in late 2004, all resided at Camp Victory near the Baghdad airport. MNSTC-I, responsible for training and equipping the ISF; the Gulf Regional Division of the Army Corps of Engineers, responsible for reconstruction project management; and Joint Contracting Command, responsible for our contracting support, all resided in the Green Zone. Task Force 6-26,* our special operations task force, resided at Balad Airbase.

We felt that we had a good plan for joining MNF-I and Embassy efforts, but it quickly became clear to the Ambassador and me that we needed to be integrated with the sovereign Iraqi government. Although developing coordination and consultation mechanisms with the new government had been specified in UNSCR 1546, bringing the Iraqi government—particularly the security leadership—into the development of a long-term, country-wide counterinsurgency effort would prove to be a daunting undertaking.

The Ambassador and I met with Prime Minister Allawi early in an informal, getting-to-know-you session. He was clear on his desire to make sovereignty as meaningful as possible considering the 162,000 foreign troops he had in his country. This would

---

* Due to security precautions, the name of the special operations task force frequently changed.

be something we would all wrestle with throughout my tenure: Iraqis rightfully wanted control, but they lacked the capacity to execute it—especially when they were trying to fight an insurgency and build their country simultaneously. I also visited the Ministry of Defense (MOD) and Ministry of Interior (MOI), and we began our first consultative sessions that week. There were two principal security forums. The Strategic Action Committee, cochaired by the National Security Advisor and me, was basically a weekly deputies committee to frame security issues for ministers. The Ministerial Committee for National Security, chaired by the prime minister and attended by key security ministers (with the Ambassador and me as invited participants), met weekly and was the decisionmaking body. We also began separate weekly dinners with the prime minister and his security ministers as confidence-building sessions to create the trust that would be essential to move forward in a common direction. It was clear to us that, even though this government was only scheduled to be in power for 6 to 7 months, it was imperative that we invest in these relationships. Looking back, I cannot overstate the importance of taking the time to build personal relationships. They proved essential in conducting the battles in Najaf and Fallujah and in preparing for the elections. Our weekly meetings and informal dinners grew in productivity over time and allowed us to stay connected with the Iraqi leadership. Our goal of One Team/One Mission was progressing and expanding to include the Iraqis.

During this first month, I began to report back to Secretary Rumsfeld on the ISF assessment that he had requested. During our many video teleconferences on the ISF, he asked some great questions: How many ISF are there really? How many did the

Iraqis really need? Did we have an effective methodology for tracking their development? How was the ISF development effort integrated into the overall strategy? Did we have agreement from the IIG on the plan? These questions guided the conduct of our assessment, which was a comprehensive effort that involved all of our major subordinate commands, and was led by LTG Petraeus. The result was a significant adjustment to the existing plan for the Iraqi security forces. The initial plan, developed in the early days of the mission, had not been based on ISF participation in counterinsurgency operations; rather, the army's three divisions would primarily provide external defense without threatening Iraq's neighbors. The police and border forces would primarily conduct community policing and peacetime operations. The situation on the ground had changed substantially since that plan was developed.

Our review incorporated an analytical look at security forces in other insurgencies and in other regional countries, and included input from our major subordinate commands on the security needs in their particular areas. The review called for finishing the existing effort and increasing the local police and border police by over 60,000 to achieve a ratio of 1 policeman to every 197 people—an acceptable planning ratio for security forces in moderate- to high-risk counterinsurgency scenarios. It also called for a significant increase in the Iraqi National Guard (a regionally recruited and trained military force that took the place of the Iraq Civil Defense Corps) by 20 battalions and adding brigade and division headquarters for more effective command and control.[6] The total cost for the expansion was almost $3 billion, but it was an essential step for our long-term success. With strong support from Ambassador Negroponte, the plan was approved by Prime Minister Allawi and the U.S. Government in mid-August.

As a result of the review, we had a new start point for the ISF that we were comfortable with, but we knew that we had a long way to go to get the Iraqis to the point where they could secure their own elections in January and ultimately take over the counterinsurgency campaign. That July, only about 30,000 of the police on duty were trained—and the new plan called for 135,000 police; only about 3,600 of the 18,000 border guards had weapons—and the new plan called for 32,000 border guards; and only 2 of the Iraqi battalions had reached an initial operating capability while the new plan called for 65.[7] We set out to build the ISF at a pace that would not only meet our operational timelines, but also ensure that the forces held together when challenged. This meant that the coalition would have to carry the security load in the near term to give the ISF time to grow and mature—we would have to fight our way to the first election.

Using the operational framework of our campaign plan, we prioritized the development of the ISF in the 15 key cities of Iraq to get them to the point where they could conduct platoon- (or police station) level operations by the first elections in January. We, and the Iraqis, felt that it was important to the legitimacy and sovereignty of the government that the Iraqis be seen as playing the predominant role in providing election security. Our longer term goal was for them to be able to secure their own country. We gradually came to appreciate the fact that building infantry battalions was the easy part, whereas creating institutional capacity* and building the entire police system were much harder. The more experience we got working with the Iraqis, the more we realized that the institutional development would take years.

Military Operations

All of this work was essential to setting the conditions for our long-term success, but the enemy did not take a break while we settled in. In early August, while the planning and assessing were going on, Sadrist forces that had been controlling and terrorizing the city of Najaf for several months attacked one of our coalition patrols. The violence escalated rapidly as Muqtada al-Sadr mobilized his forces in Baghdad and the southern part of the country. This provided a key opportunity for the prime minister and the new government to demonstrate their strength by restoring Iraqi government control to Najaf. As the elimination of the Sadrist stranglehold on the population of Najaf would need to be seen as a largely Iraqi operation and there were very few capable ISF available, success required careful integration of the political and military efforts. With some masterful tactical actions by coalition forces and Iraqi commandos that involved brutal hand-to-hand fighting, some careful management of ISF coming directly out of training into the operation, and excellent political-military interaction at all levels, we managed to evict the Sadrists first from their base in the cemetery and finally from the city itself. Al-Sadr was left no option but to negotiate his personal release. The IIG had its first victory.

---

* For example, building the training, education, intelligence, financial, administrative, logistical, and procurement institutions.

ESTABLISHING THE MISSION

We followed up the military operations with quick-impact construction projects to demonstrate to the Iraqi people that there was a benefit to supporting their government. While this part of the operation took months and did not have the immediate impact we had hoped, it did help consolidate our gains. Thus the "Najaf Model" was born: hard-hitting coalition-Iraqi military operations, with political support mustered by the Iraqi government, followed by focused reconstruction efforts. This model later formed the basis for the "clear-hold-build" concept.

My relationship with Prime Minister Allawi was critical during this battle as we melded coalition military power with the legitimacy of his government. He emerged from the battle for Najaf more confident and focused upon building momentum toward resolving the security situation across all of Iraq. As we looked to learn from what happened in Najaf, we gained many insights, to include legitimizing coalition military operations with the help of the Iraqi government, incorporating even small numbers of ISF in military operations to put an Iraqi face on these operations, and integrating reconstruction projects immediately following military operations. Most importantly, Prime Minister Allawi had emerged as a leader, and he was gaining the confidence of the Iraqi people.

Building upon this operational success, I urged Prime Minister Allawi to adopt a country-wide strategy to eliminate the terrorist safe havens across Iraq before the elections. I suggested that we conduct joint operations to neutralize safe havens in Baghdad (Sadr City), Samarra, and Fallujah. I gained the prime minister's concurrence on this approach in September, setting the operational agenda for the next several months as we turned our focus to the January

STRATEGIC REFLECTIONS

2005 elections. We were constrained in this effort by the limited number of ISF and the pace at which we could build them if we expected them to hold together. Over the next several months, we worked with the Allawi government to reduce Sadr City and Samarra as terrorist safe havens. Last up was Fallujah.

As we executed this strategy, whether we really needed to eliminate the terrorist safe haven in Fallujah became a difficult policy question for the prime minister. Prime Minister Allawi, the Ambassador, and I wrestled with this until we became convinced that having a terrorist safe haven within 30 miles of the capital presented an unacceptable risk to the conduct of the upcoming election. The specter of the April "failure" in Fallujah hung over the discussions, and the fact that we would be attacking a largely Sunni area at a time when we were trying to bring the Sunni population into the political process weighed heavily on us. The Ambassador strongly supported the operation, and we told the prime minister that if we undertook the operation, we would be successful, but that there was a high risk of coalition and Iraqi casualties and collateral damage. We also told him that if we started it, we had to stay at it until we finished. "Start together, stay together, finish together" became our motto. To his credit, Prime Minister Allawi not only accepted this, but he also built and sustained the Iraqi political support that allowed us to complete the operation. Fallujah was the toughest tactical battle of my time there, but more than any other operation, it opened the door for the successful elections in January.

Election Planning and Preparation

Concurrent with this operational focus, we began the complex planning for the first free elections in Iraq since 1954. The

Independent Iraqi Election Commission had the lead with advisory support from the United Nations. The Iraqis wanted to provide the security, or at least be seen as providing the security, but would still need our logistical and security support. This tension in the planning effort between what the Iraqis wanted to do and what they were capable of doing was constant throughout my tenure, but most pronounced during this period because of the strategic importance of the elections. We constantly struggled with finding a balance between putting an Iraqi face on important events and not allowing them to fail. Finding this balance was more of an art than a science.

From a military perspective, we set up a series of offensive operations in the aftermath of Fallujah to maintain the momentum and to keep the pressure on the insurgents. To do this, I requested 10,000 additional troops to get us through the elections. I also personally visited every province to check on election preparations. It became clear to me that, as chaotic as things might look from Baghdad, coalition commanders and Iraqi provincial leaders had a good handle on things at their levels. It began to appear that we were positioned to execute successful elections, which we defined as fairly conducted elections in which every Iraqi who desired had the opportunity to vote. We knew that there would be violence as the insurgents worked to unhinge the elections, but I felt we had a good plan—further enhanced by some last minute curfews and movement restrictions—and that we would be successful.

On the political side, a looming election issue was Sunni participation—would they boycott the election and, if so, what would be the post-election implications? The Ambassador and his staff worked this very hard, and our commanders in the field reinforced their efforts with the Sunni populations in their areas of operation. Ultimately, a good number of the Sunni population did boycott the election*—an occurrence that would have long-term implications for our mission—but the elections were still largely successful from a security perspective.

Looking Beyond the Elections

After the Fallujah operation in November, when the progress of military operations and the growth of the ISF made it increasingly likely that the January elections could be held on time and with an Iraqi face to enhance the legitimacy of the IIG, we began looking ahead to 2005. We drew on the experience that we had gained in 5 months on the ground and on several key studies and assessments to formulate our plans.

In the early fall of 2004, I directed my planners to review counterinsurgency best practices to see what we could learn from history. "COIN" was something that we, in the U.S. military, had not been involved with for some time. My perception, from observing and talking to subordinates, was that we understood the doctrine well enough, but that we all had a lot to learn about how to apply that doctrine, particularly in Iraq. The staff did their usual good work and examined a series of 20th-century counterinsurgency campaigns for successful and unsuccessful practices (see figure 2-5). They developed a comprehensive report that included the list below that I shared with the Secretary of Defense,

---

* Turnout of Sunni voters was as low as 2 percent in Anbar Province.

Figure 2-5. Counterinsurgency Practices

SUCCESSFUL (average 9 years)	UNSUCCESSFUL (average 13 years)
Emphasis on intelligence	Inferior intelligence operations
Focus on population, their needs & security	Primacy of military direction of counterinsurgency
Secure areas established, expanded	Priority on "kill-capture" of enemy, not on engaging population
Insurgents isolated from population (population control)	Ineffective, minimal psychological operations campaigns
Unity of Effort on multiple lines of operations, local to national	Battalion-size operations as the norm
Resources (money, manpower, time)	Military units concentrated on large bases for protection
Effective, pervasive psychological operations campaigns	Special Forces focused on raiding
Amnesty & rehabilitation for insurgents	Advisor effort a low priority
Police in lead, military supports	Building, training indigenous army in image of own (foreign) force
Police force expanded, diversified	Peacetime government processes
Conventional military forces reoriented for counterinsurgency	Open borders, airspace, coastlines
Special Forces, advisers present with indigenous forces	
Insurgent sanctuaries denied	

the Chairman, Service chiefs, and my subordinate commanders in Iraq. The two key insights that I drew from the report were the average successful COIN campaign in the 20th century lasted 9 years (unsuccessful ones lasted longer), and no outside power had been successful in prosecuting an insurgency without a capable indigenous partner.[8]

Additionally, in early December, the Ambassador and I directed another U.S. Embassy/MNF-I Red Team to make an independent assessment of possible election outcomes. The team projected three election result possibilities: a two-thirds Shia majority (which meant that the Shia parties would control the parliament), a 50 percent Shia majority (which meant that the Shia parties would have to negotiate with other parties to form a government), or total disruption of elections. While the team did not believe that the last outcome was likely, it did warn that major political and security challenges would await coalition forces in both cases of a Shia majority. Furthermore, the team concluded that there were no near-term prospects that insurgency and terrorist violence would lessen; rather, the Sunni-based insurgency might grow more intense when confronted with a strong Shia-governing majority. The aftermath of the election later revealed how prescient these predictions were.

Also in December, the MNF-I staff conducted its first Campaign Progress Review (CPR). By this time, we had settled on bimonthly assessments to monitor progress and drive campaign adjustments. While useful, I felt that the 2-month horizon was too short for longer term planning and that we periodically needed a broader assessment of the campaign. We decided to measure our progress at 6-month intervals based on the six strategic effects from the campaign plan: a legitimate Iraqi government, neutralization of the insurgents and terrorists, a capable ISF, basic Iraqi needs met, a wedge driven between the insurgents and the people, and changing Iraqi perceptions of the coalition. The December CPR marked the first formal semiannual assessment. I made it a point not to overly involve myself in the production of the document, and found it a good way to find out what "The Colonels" were really thinking. This candor was further enhanced because the drafter, as a result of coalition-agreed staffing, was always a British colonel, who soon gained the moniker "the Gloomy Brit."

The staff concluded in the December 2004 CPR that our strategy was "sound, but must be implemented more effectively to succeed—particularly along the non-kinetic lines of operation" and that the "Iraqis must play an increasingly larger role."[9] They objectively looked at what we had accomplished and gave their assessment of where we needed to focus our efforts in the year ahead. They felt that some things had gone well:

- elimination of terrorist safe havens and the suppression of the Shia insurgency
- improvement in IIG capacity
- election planning was broadly on track
- the pace of reconstruction was accelerating
- generation of ISF was exceeding MNSTC-I projections (more than 20 percent growth since August)
- the coalition was still largely intact.

On the other hand, the review noted that we still had a number of challenges:

- an intensifying insurgency of former regime elements (who would not be defeated by military means alone)
- police and border guard capacity was still particularly weak and needed to be improved
- the engagement program with Sunni Arabs was limited and needed to expand
- ISF operational performance was mixed, but generally more positive in the presence of coalition assistance teams
- economic development and communication lines of operation were not performing at full potential
- building unity of effort between MNF-I, the Embassy, and the Iraqi government needed work.

Overall, I was pleased with where we had gotten in 5 months, but it was becoming increasingly clear to the Ambassador and me that building the capacity of the Iraqi government to a minimally acceptable level, particularly the ISF, was going to take a lot longer than the 18 months covered by UNSCR 1546 and our campaign plan, especially if there were going to be two changes of the Iraqi government in that period.

Armed with these insights, I returned to Washington in mid-December for consultations on the situation in Iraq and to provide my thoughts on the way ahead after the January elections. I met with the President, Secretary Rumsfeld, and the Joint Chiefs and shared the findings of our Campaign Progress Review, COIN, and Red Team studies, and reported on our preparations for the upcoming elections. I also began to discuss the concept of placing coalition advisor teams alongside Iraqi military, police, and border forces to hasten the development of our "capable indigenous partner." I told the national security team that we would be ready to conduct the January elections, but that there would be violence as insurgents and terrorists attempted to disrupt the elections. I also warned them to expect a loss of momentum during the government formation process after the elections, and—to emphasize our thinking on how long this might take—I stated that even if the UNSCR 1546 process was completed on schedule, the Iraqis would still face an insurgency, long-term development challenges, and meddling neighbors. I also pointed out that a year from then, the ISF would still not be capable of independent COIN operations.[10] My message was that the mission in Iraq was going to extend beyond the 18 months of the UN timeline, but we would be ready for the first democratic elections in over five decades.

The First Elections: January 30, 2005

Upon my return to Iraq, I focused on the execution of the elections. With the additional troops that I had requested for election security, we had kept pressure on the insurgents throughout December and January. Keeping the insurgents off balance allowed us to focus on securing the election process and ensuring that the ISF would be seen as the face of election security. Although we had been working hard, and had been largely successful in achieving local control (ISF capable of platoon- or police station–level operations) in the 15 key cities of Iraq, the ISF still were not ready for a mission of this scope—delivering and recovering election material to and from 5,200 polling stations across Iraq and securing them. Together we developed an "inner ring/outer ring" plan to secure the polling stations to prevent a determined effort by insurgents and terrorists to stop or significantly disrupt the elections. The ISF would secure the inner ring (the area immediately surrounding the polling stations), and coalition forces would secure the outer ring (the approaches to the cities and polling stations). I personally visited all provinces in the weeks prior to the election to ensure this concept was understood, and also supervised a "ROC drill"—an election-day rehearsal—with key Iraqi, Embassy, and coalition leaders. We were as ready as we could be.

On January 30, our detailed preparations paid off as over 8 million Iraqis—58 percent of the eligible population—turned out to vote. There were almost 300 attacks on election day, but our operations and ISF security of the 5,200 polling stations ensured that insurgents did not significantly disrupt the voting. The Iraqi people had freely elected their parliament over the course of what was a very emotional day in Iraq. President Bush spoke to Prime Minister Allawi and congratulated him, and also addressed the American public that evening in a televised speech. MNF-I had begun 2005 by achieving what 6 months before President Bush had announced as his "most important" task. It was a good feeling.

But as things went in Iraq, we had to take the bad with the good. As pleased as we were with the election turnout and security efforts, the lack of Sunni participation meant that they would have limited influence in the development of the constitution, and this did not bode well for our efforts to defeat the insurgency.

U.S. Ambassador to Iraq John Negroponte (left) greets Interim Iraqi Government Prime Minister Ayad Allawi

3. COMPLETING THE UN TIMELINE AND SETTING CONDITIONS FOR A NEW IRAQ (JANUARY–DECEMBER 2005)

Perhaps the best insight into what I was thinking as we entered 2005 can be gained from the first several paragraphs of an assessment I wrote to my boss, General Abizaid, on January 5, laying out my plans for 2005. I began that with the famous passage from T.E. Lawrence that provided advice on dealing with Arabs: "Do not try to do too much with your own hands. Better the Arabs do it tolerably than you do it perfectly. It is their war, and you are to help them, not to win it for them."[1] It was a mindset that I had concluded we would need to instill in coalition forces if we were to be successful in Iraq. I told General Abizaid:

51

I shared our 5 month assessment with you in DC. We used it to get a comprehensive view of where we stand in executing the Campaign Plan, and to frame our thinking for an approach for 2005. We believe that we are broadly on track. We have rolled back insurgent gains and eliminated insurgent and terrorist safe havens in Iraq, suppressed the Shia insurgency, quintupled reconstruction activity, kept ISF development roughly on track, made progress in local control in 14 of 18 provinces, and saw the growth of Iraqi governmental capacity. Election preparations are proceeding in all but Ninewa and Al Anbar provinces.

We have also seen a [former regime element] *insurgency that has gotten better organized, that is conducting a campaign of intimidation in the Sunni areas that threatens to unhinge political, economic and security force development, and that is creating a real sense of uneasiness about the security situation for the upcoming elections. Further, while ISF development has progressed, they still lack the capacity for independent action, absenteeism threatens the viability of our training and equipping programs, and Iraqi intelligence organizations have not developed as hoped.*

That said, our objective of an Iraq "at peace with its neighbors and an ally in the War on Terror, with a representative government that respects the human rights of all Iraqis, and security forces sufficient to maintain domestic order and to deny Iraq as a safe haven for terrorists" is still attainable. However, security force development won't be completed in '05, and sustaining our investment will take even longer.[2]

I strived hard to achieve balance in my reporting. Things in Iraq were never all good or all bad, so I tried to highlight both the positive and negative aspects of the situation. In a mission as complex as Iraq, you must make slow, steady progress even as you deal with setbacks—and setbacks are a reality of war. We were positioned for successful elections, but we had a long road ahead of us.

A number of things had also become clearer to us after 6 months on the ground. First of all, it was going to take much longer than the 18 months of the UNSCR 1546 timeline to complete our mission. Ambassador Negroponte, in a December 2004 cable to the Secretary of State, suggested we should be thinking in terms of at least a 5-year time horizon. Our counterinsurgency study had found that successful insurgencies had historically lasted around 9 years. There was no reason to think Iraq would be different. During my December meetings in Washington, I had emphasized that resolving the situation in Iraq would take longer than the 18 months of the UNSCR timeline. We clearly needed to expand our horizons beyond the end of 2005.

Second, the Shia political parties and politicians, who were going to lead the Iraqi Transitional Government, had little experience in governing. Their election represented a reversal of the governing situation that had prevailed in Iraq for the last three and a half decades. We were concerned about this from two perspectives: inexperienced ministerial leadership would inhibit ministerial capacity-building, and the reversal of the governing situation would feed the feelings of disenfranchisement in the Sunni population that former regime elements had been leveraging to sustain the insurgency. We were further concerned about the limited Sunni turnout in the elections, which meant that this population would be underrepresented in the drafting of the constitution.

Third, while the training and equipping of the ISF were generally on track, it was going to take much longer to get them to the point where they could credibly and independently assume the lead of a nationwide counterinsurgency campaign. If we wanted this to go faster, we would have to commit more resources and look at new approaches to their training.

Fourth, in spite of our best efforts to improve the Iraqi view of the coalition, much of the population still viewed it as an occupation force, and while there was a clear understanding at the governmental level that we intended to leave, there was apprehension at the local level that we were in Iraq to stay—an apprehension that was manipulated by the insurgents. This tension of having a large foreign Western force in a sovereign Middle Eastern country was a constant friction. We needed to demonstrate that we had a plan to leave.

Fifth, we expected that there would be a loss of momentum in the period following the election as the transitional government was selected, formed, and transitioned into the job. We looked for ways to sustain the momentum of the elections through this period but were largely unsuccessful outside the security sector.

Finally, 2005 was to be a year of key transitions: there would be two transitions in the Iraqi government (the last one spilling over into 2006) bringing with them changes in ministers and other key personnel; MNC-I and its subordinate units would change in February; and the key leaders in the Embassy and MNF-I, who had done such great work to get us to this point, would change out in the summer, taking with them the wealth of experience gained.* The turbulence generated

---

* At that time, we did not know about Ambassador Negroponte's March departure to become the first Director of National Intelligence.

as new people and units came and went was a significant complicating factor, particularly on the Iraqi side. We would have to work hard to mitigate the impacts of the frequent transitions.

Adapting in Transition

The transitions at the political level significantly impacted our strategic momentum. While we were able to mitigate the loss of operational momentum caused by the MNC-I transition, political momentum proved tougher to deal with. In retrospect, we underestimated the impact of three government transitions in 2 years on our ability to build capacity in the Iraqi ministries and to provide consistent leadership to the Iraqi people. The governments were just starting to get a feel for governing when they were replaced. Prime Minister Allawi held office for 11 months (4 months of which were a "lame duck" period following the January 2005 elections). Ibrahim al-Jafari was prime minister for 13 months (6 months of which were lame duck). We had an initial view that we could continue to build ministerial capacity through the transitions. This did not turn out to be the case. In fact, we lost ground in many ministries during the al-Jafari government. The small pool from which to draw qualified ministers and the lack of an established government bureaucracy meant that we had to almost start from scratch with each new government and minister. What would continue to become clearer was that, in these types of operations, everything takes longer than you think—particularly those things over which you do not exercise direct control.

Armed with these insights, we began 2005 certainly wiser than we had been when we arrived 6 months before, but with a full plate of very difficult issues. The Ambassador and I set out to adjust our strategy and plans for 2005 based on these insights. We did this during the transition period of the new government to posture ourselves to move forward once the ITG was seated.

Building a Transition Concept. Our review of our plans began with the Campaign Progress Review in December, which helped shape the insights mentioned above and focused us on what we needed to accomplish in 2005.

As we looked ahead, we wrestled with the realization that we could well complete the UN timeline on schedule by the end of the year, but that Iraqi government and security forces capacity would not be at a point where Iraqis could credibly take responsibility for their own security and governance as UNSCR 1546 had envisioned. At that time, we had not been through the actual transition of governments, so we made this judgment with the projection that ministerial capacity would continue to grow as governments transitioned—an assumption that did not pan out. We knew that our mission was ultimately to hand over security responsibility to the Iraqi government, but we had not yet developed a concept to do this. We thought that if we could demonstrate a plan to build credible Iraqi security capacity as rapidly as possible and follow that with a conditions-based plan to transition the security mission to the sovereign Iraqi government, we would come closer to meeting the expectations of Washington, the Iraqi government, and the Iraqi people. We were very cognizant of the fact that we would need to continue our efforts to defeat terrorists and insurgents while executing this concept.

The Ambassador and I worked together to shape this thinking into a second joint mission statement that we issued to our subordinates on February 7, shortly after the elections. In it we stated, "In 2005, we will work closely with the Iraqi Transitional

Government . . . to diminish the insurgency and prepare the Iraqi Security Forces and the ITG to *begin* to accept the counterinsurgency lead. We also will support the ITG's efforts to complete the timetable laid out in UNSCR 1546."[3] These two sentences captured our major missions for 2005, but it was the last two sentences of the overview that captured a change in mindset that would be essential in accomplishing this strategy: "We must always remember that we have transferred sovereignty to the Iraqis; they have elected a Transitional Government; and they will begin to take the counterinsurgency lead. There is a consistent message here: Iraq's destiny belongs to Iraqis; they want to control it; and the more they do for themselves, the more they will value the results."[4]

Inculcating this mindset into the coalition forces and Embassy staff would prove difficult, as we were pressing against the "can-do" culture of two high-performing organizations. Things were hard enough to get done in Iraq, but they were easier if we did them ourselves. Helping the Iraqis help themselves would be more difficult and take longer, but it would get us to our objectives faster. We had to discipline ourselves to build for the future while we dealt with the very difficult present.

The work underpinning this adaptation to our plans had been ongoing since November. Following the Fallujah operation, we noted that the performance of Iraqi units with embedded coalition advisors was far superior to those without them. We also found that we had much better accountability of the weapons and equipment that we gave the Iraqis with advisors present. We had substantially increased the size of the ISF—forming more than 80 army and special police battalions by February 2005—and some of these forces were at a point in their development where they could benefit from more coalition experience and expertise.

Given my personal experience in the Balkans, I was concerned that the longer we waited to begin giving Iraqis responsibility for their own security, the more dependent they would become on us—and the longer we would remain in Iraq. We came to believe that by embedding coalition advisors with Iraqi military, police, and border police units, and by aligning Iraqi units in supportive relationships with coalition units (called "partnership"), we could not only accelerate their development, but also get them more actively involved in the counterinsurgency fight sooner—increasing the forces we could field against the insurgents. I directed my staff to begin working the details of how this might work and alerted the incoming corps to be prepared to implement this new approach on arrival.

As this approach would require some 2,500 additional U.S. forces and the approval of the Iraqi government, I began discussing it with coalition and Iraqi leaders. Washington was concerned about the safety of the teams living and working with Iraqi units, the impact that pulling 2,500 officers and senior noncommissioned officers out of units would have on the Services, and the naming of the teams (*transition* teams was chosen over *advisor* or *assistance* teams to highlight that these teams were part of a process to "transition" security responsibility to the Iraqis). The Iraqi leaders saw value in accelerating the development of the ISF, but wanted the program implemented in a way that did not impact on their sovereignty. (Prime Minister Allawi would not agree to police transition teams working in local police stations for this reason.) Washington approved the additional forces in March, and we signed memorandums of understanding with the outgoing ministers of interior and defense in April to implement the transition team and partnership programs. We expected that

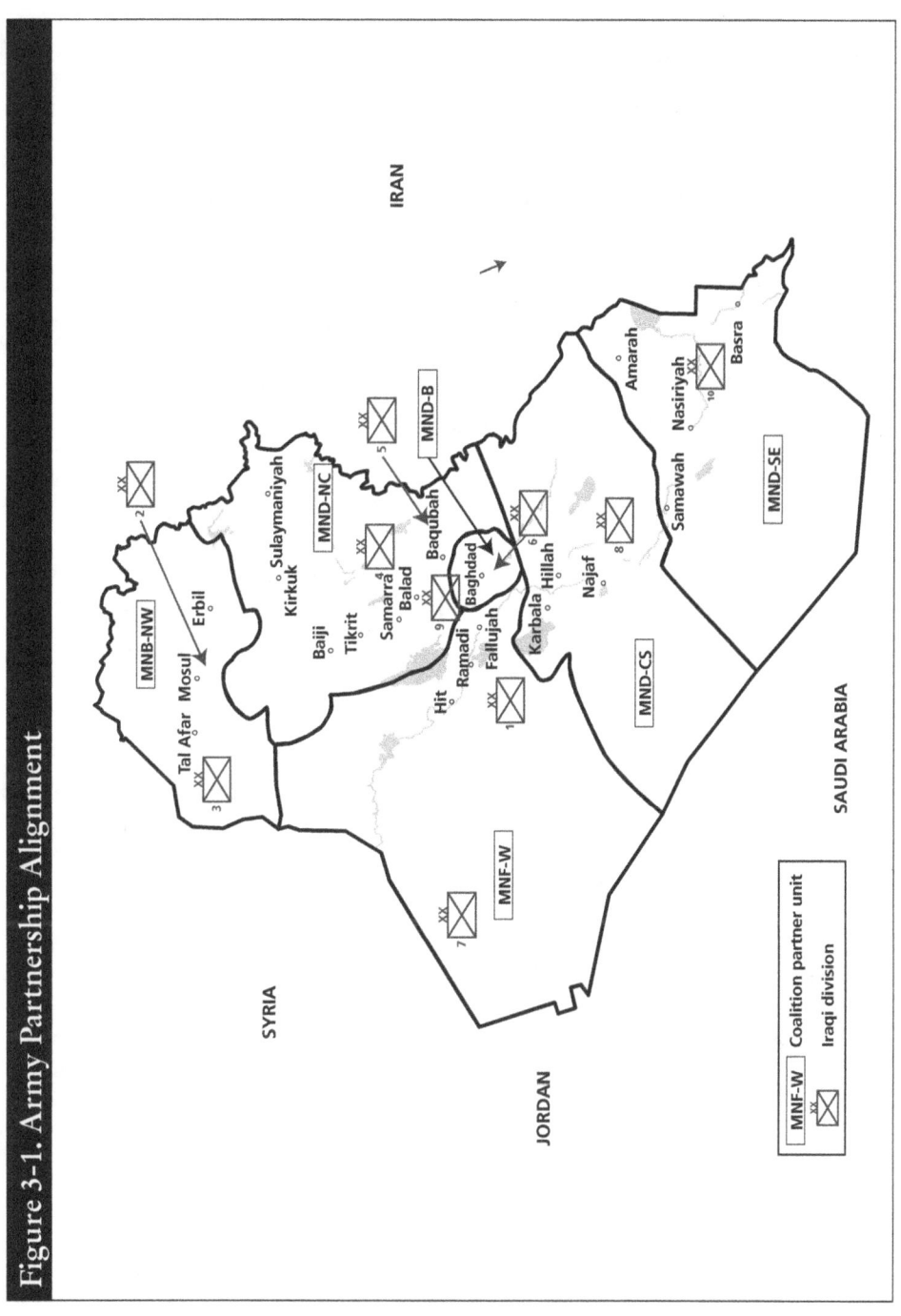

Figure 3-1. Army Partnership Alignment

STRATEGIC REFLECTIONS

we would have to revisit the agreements with the incoming government and ministers, but felt that it was important for continuity to get the existing government on board—in writing. Our plan was to establish partnerships and embed the teams by June, so Iraqi government commitment and direction to accept the program was essential. As seen in figure 3-1, we planned to align our divisions with every Iraqi division (and special police brigade) and establish partnerships down to battalion level to facilitate the interaction between Iraqi and coalition units.

Getting the transition team and partnership programs in place to accelerate ISF development was the first step in the process of transitioning security responsibility to a capable Iraqi military. We envisioned that as the ISF became capable of conducting counterinsurgency operations, first with transition teams and coalition-enabling forces (for example, close air, artillery, and logistic support), and ultimately independently, coalition forces would transition to a supporting role and gradually reduce their presence. All transitions of security responsibility were to be based on the readiness of the ISF to conduct counterinsurgency operations, so we devised and implemented a readiness reporting system with the Iraqis based on that criterion to track the development of their forces. We envisioned a four-phase plan for the transition of security responsibility:

- ◆ Phase I: Implement the transition team concept
- ◆ Phase II: Transition to provincial Iraqi security control (Iraqis in charge of security at the provincial level, with coalition support)
- ◆ Phase III: Transition to national Iraqi control (Iraqis in charge of security at the national level, with coalition support)

- Phase IV: Iraqi security self-reliance (Iraqis in charge of their own security).

It would be summer before the new Iraqi government was ready to seriously discuss this process. This original plan was adjusted as a part of those discussions. We did not set specific timelines for any phase but Phase I (June 15) as we expected all transitions to be conditions-based.

Changing Our Mindset. We were very conscious that this adjustment to the campaign plan entailed a major shift for our conventional Army and Marine forces. For the first time since Vietnam, we would be asking conventional forces to actively participate in the training of indigenous forces during combat operations. To make this work, we recognized that we would not only have to "train the trainers"—that is, to teach conventional forces the art of training and working with indigenous forces—but we would also have to change the mindset of our forces away from doing things themselves to helping the Iraqis do them.

To train the trainers, we established an in-country training center, Phoenix Academy, and used our Special Forces to educate our conventional forces. To ensure the concept was understood across MNF-I, the staff produced a campaign action plan for 2005 that captured the adaptation to our plan and adjusted our mission accordingly. Our endstate remained the same. We wanted to progressively shift the coalition main effort from fighting the counterinsurgency ourselves to transitioning the responsibility for fighting the counterinsurgency to the Iraqi government and security forces. We would accomplish this by increasing our capacity to improve ISF capabilities—transition teams and partnerships—and by conducting aggressive counterinsurgency operations to bring the insurgency to levels that could be contained by increasingly capable ISF. The revised mission statement reflected this strategy:

> *In partnership with the Iraqi Transitional Government, MNF-I progressively transitions the counterinsurgency campaign to the ITG and Iraqi Security Forces, while aggressively executing counterinsurgency operations, to create a security environment that permits the completion of the UNSCR 1546 process and the sustainment of political and economic development.*[5]

As I went around the country to receive unit backbriefs on the action plan, I reinforced the fact that we would *progressively* implement this mission because all transitions would be based on ISF capabilities and those would vary widely across the country. I encouraged leaders to be candid in their assessments of the ISF because I would make some difficult decisions based on those assessments. I also met with the commanders of each incoming unit within 30 days of their arrival to personally brief them on the concept and to answer their questions directly.

Finally, to communicate directly with the coalition servicemembers, I issued a set of "flat-assed rules" (FARs, see figure 3-2) to every incoming coalition service man and woman. The FARs were intended to convey my priorities for success in Iraq and to instill the mindset that we were there to help the Iraqis restore control to their country, not to do it for them. By the late spring, the time when the transitional government was finally appointed, we were well on our way to implementing the new approach, but the change in the mindset would take a lot longer.

> **Figure 3-2. Commanding General's FARs, February 2005**
>
> - Make security and safety your first priorities.
> - Help the Iraqis win, don't win it for them.
> - Treat the Iraqi people with dignity and respect. Learn and respect Iraqi customs and culture.
> - Maintain strict standards and iron discipline every day. Risk assess every mission—no complacency!
> - Information saves lives—share it and protect it.
> - Maintain your situational awareness at all times—this can be an unforgiving environment.
> - Take care of your equipment—it will take care of you.
> - Innovate and adapt—situations here don't lend themselves to cookie-cutter solutions.
> - Focus on the enemy and be opportunistic.
> - Be patient. Don't rush to failure.
> - Take care of yourself and take care of each other.

Implementing the new mission was not an easy task for the MNC-I commander, LTG John Vines. The U.S. military was still adapting to counterinsurgency operations and now he had a new mission to develop indigenous forces while keeping the pressure on the terrorists and insurgents. That he had to do this in a political vacuum as the new government was forming made his task even more difficult. While MNC-I was transitioning to the new mission, it was conducting counterinsurgency operations across the country to keep pressure on the terrorists and insurgents and executing the operational tasks from the campaign plan: neutralizing the insurgency in the Sunni Triangle, securing Baghdad and the borders, sustaining support in Shia and Kurdish areas, and providing security for the October constitutional referendum and the December elections. I received a backbrief from LTG Vines early in his tenure on how he intended to accomplish these tasks. Following that, with few exceptions, I left the day-to-day management of these operations to him. The exceptions involved operations that I felt were critical to the theater mission. In 2005, this meant MNF-directed operations to restore Iraqi control to their western border, operations that I discuss later.

Keeping Washington Up to Date. Shortly after Prime Minister al-Jafari was appointed and the new government was formed in early May, we conducted our second semiannual Campaign Progress Review.[6] While this was intended to be a 6-month assessment, it became an assessment of the campaign over the last year.

Armed with this assessment and a year's experience on the ground, General Abizaid and I visited Washington in June. Theater commanders have a role in helping the administration communicate about their mission to the American people and Congress. Given that I had been on the ground a year and that we had made some substantial adjustments to the mission, it made sense to go back and update Congress and national security leadership on the mission and to conduct engagements with the U.S. media. I returned to Washington three times in 2005 and a total of seven times in my 32-month tenure.

I reported to the President, Secretary of Defense, and Joint Chiefs that the campaign was broadly on track and that the transition team and partnership programs were already paying dividends with the Iraqi security forces—but that they remained 18 to 24 months from conducting independent counterinsurgency operations. I noted that the insurgency remained active and would contest the completion of the UN timeline and that Iraq would still face an insurgency, meddling from unsupportive neighbors, immature security forces, and long-term political and economic development challenges after that timeline was completed at the end of 2005. This would, I reported, require a sizable coalition presence in 2006 in Iraq. I also suggested that with the improvements we were already seeing in the ISF as a result of our increased focus, and the potential for that to continue, we should begin planning to reduce coalition presence in Iraq over the next 2 years as the ISF became more capable. I felt that it was important for both the coalition and Iraqi government to be seen as honoring our pledge to pass security responsibility to the Iraqis as they were capable of accepting it. I did caution that we should not establish timetables for coalition withdrawal because the situation was still too fluid and the insurgents and terrorists would use those timelines against us. I had concluded by that time that we would not ultimately be successful in Iraq until we had brought the insurgency to levels that could be contained by increasingly capable ISF, passed security responsibility to them, and departed Iraq.

In my discussions with Congress, I tried to emphasize that, while our work in Iraq was hard and the environment was challenging, we made progress every day. I emphasized what we and the Iraqis had accomplished in the past year: the establishment of the interim government and peaceful passage of power to the transitional government, elimination of terrorist safe havens across Iraq, neutralization of the Shia insurgency, qualitative and quantitative increases in Iraqi security forces, and increased pace of economic development. To refute the notion that terrorists and insurgents had the upper hand, I also emphasized what the insurgents and terrorists had *not* accomplished: they were unable to reconstitute their safe haven in Fallujah, they had not expanded their base of support (the insurgency remained largely confined to 4 out of 18 provinces), they had not prevented the growth of Iraqi security forces (then around 170,000), they had not yet sparked sectarian violence, and they had not stopped political and economic development. I told Congress that after a year on the ground, I felt that the mission was "both realistic and achievable," but that it would require patience and will for us to succeed.[7]

While I was back, I was asked by the President to continue to serve as the MNF-I commander for another year. When I left for Iraq the previous June, I was told to plan to be in Iraq for 12 months, but it quickly became apparent that staying through the completion of the UN timeline in 2005 made more sense. He was asking me to stay beyond that until June 2006, and I agreed. I took a week of leave with my family to recharge my batteries for an even longer haul.

Completing the UN Timeline

By June, we had made good headway implementing the transition team and partnership program, and we had begun to build our relationships with the new prime minister and his security ministers. Again, I felt it very important to invest in these relationships even though we expected the new team to be in place for less than a year. Our focus for the rest of 2005 was to continue building the Iraqi security forces, governance, and economic development while we established a security situation that would allow the completion of the UNSCR 1546 timeline—a constitution by August, a national referendum on the constitution by October, and elections for a government based on that constitution by the end of the year. It also involved planning for the post-election period. We would accomplish this while bringing on board a new U.S. Ambassador, Zalmay Khalilzad, who arrived in late June from Afghanistan, and a new Iraqi government.

I viewed my relationship with the Ambassador as my most important relationship, and we spent quite a bit of time together discussing how he saw the mission and the way ahead. He agreed immediately on the One Team/One Mission concept, and was even willing to look at ways to take the integration of our efforts further. We also formed another Red Team to look at the nature of the enemy and the war and to give us their thoughts on how we were executing the mission. The Red Team was again under U.S. Embassy leadership with MNF, CIA, and UK participation. The team looked at how to achieve decisive results by prioritizing and synchronizing our finite resources to "break the back" of the insurgency in 1 year and to defeat the insurgency in 3 years, and it also made some useful suggestions on how to better integrate our counterinsurgency efforts.

This independent assessment, along with our recently completed June Campaign Progress Review, helped us shape a common vision for the way ahead. We began to plan in earnest for the mission to continue beyond the end of the year—the conclusion of the UN mandate.

As we looked beyond the December 2005 elections, we were again concerned with the potential loss of momentum because of the political and operational transitions in early 2006 (we would also face a transition of MNC-I during the government transition and formation period), so we redoubled our efforts to leverage the benefits of completing the UN timeline and seating an Iraqi government based on an Iraqi constitution. Given the uncertainty of the December election outcome and what lay beyond the end of the year, we developed interim guidance to leverage the momentum of successful elections while we worked on a broader campaign plan for the post-election period.

This interim guidance became known as the "Bridging Strategy" because it bridged the gap between the fall of 2005 and the seating of the constitutionally elected government that would come sometime after the first of the year. The strategy was focused on setting the conditions for the December elections and shaping the aftermath in ways that would have a decisive, positive impact on our mission. To be decisive, we felt that we had to defeat al Qaeda in Iraq (AQI) while bringing and keeping the Sunni population in the political process in a manner that began to neutralize the insurgency. We had to do both of these tasks while continuing to grow and develop ISF capabilities and, as these capabilities grew, by placing the Iraqis increasingly in the lead of security operations. "Al Qaeda out, Sunni in, ISF in the lead" became the shorthand version of the strategy, and it drove our efforts in the second half of 2005 and into the spring of 2006.[8] Concurrently, the Ambassador and I directed the development of a new joint campaign plan, which would be the first plan to be jointly prepared with full Embassy integration in the planning process. It would cover the 4-year tenure of the new Iraqi government.

With the new Ambassador also came a renewed interest in Sunni engagement and Provincial Reconstruction Teams (PRTs). Sunni engagement was something that General Abizaid and I had been pushing for some time. We had pressed hard for an engagement strategy with the Sunni population as a means of driving a wedge between the general Sunni population and the terrorists and insurgents. Early in the mission there was reluctance to do this for fear of alienating the Shia population that had been disenfranchised for so long. In my view, we would not make progress with the Sunni population, and, as a result, the insurgency, unless we facilitated a dialogue to bring them into the political process. While our subordinate leaders were engaging tribal leaders locally, there was no cohesive mission effort to do this. With the arrival of Ambassador Khalilzad, we began a concerted effort to bring the Sunni population into the political process. This effort set the conditions for increased Sunni participation in the constitutional referendum and December elections, and for continued post-election dialogue, particularly in Anbar Province.

We had begun a pilot effort in the summer of 2005 to establish seven Provincial Support Teams based on the PRT model that was developed in Afghanistan during Ambassador Khalilzad's time there. (We actually sent a team to Afghanistan that March to see how their PRTs functioned.) We allocated $70 million from the Commander's Emergency Response Program to fund development and reconstruction in these provinces. Our intent was to bring coalition support to the provincial level to facilitate development from the bottom up. It had become clear that the top-down approach from Baghdad was not working. With the Ambassador's arrival, we enhanced this effort in the fall by appreciably increasing the size of three PRTs in the key provinces of Mosul, Kirkuk, and Babil. After considerable interagency discussion with Washington, we received approval to go forward with enhancing the remaining PRTs in 2006. Although they began to have a positive effect almost immediately, their impact was inconsistent—in my view, too tied to the personal initiative and competence of the PRT leader. It would take a while to even out the performance, but, over time, they accomplished their intended effect.

STRATEGIC REFLECTIONS

Military Operations. With the seating of the al-Jafari government in May, we began to see an increase in suicide attacks focused primarily on the Shia civilian population. These attacks had the potential to unhinge the progress we were making on the political side by exploiting existing Sunni/Shia tensions. Our intelligence analysts began to see a shift away from the Sunni insurgency as the most dangerous threat to the accomplishment of our mission to the Islamic extremists who were conducting the attacks. We became concerned that if we did not reduce the ability of the extremists to conduct suicide attacks across Iraq, the coming constitutional referendum and December elections could be in jeopardy. Our analysts believed that the vast majority of suicide bombers were not Iraqi and entered into the country by crossing the Syrian border. They were moved to their targets by facilitation networks along the western Euphrates valley and Tal Afar–Mosul corridor. Accordingly, I directed the MNC-I to conduct operations to defeat those networks and restore Iraqi control to the borders before the December elections. This would become the major MNC-I operational focus in the runup to the elections as it also continued to focus on securing Baghdad, steady-state counterinsurgency operations across Iraq, and developing the Iraqi security forces.

The operations in the west required careful integration of the actions of our special operations task force, which was targeting the al Qaeda leaders of the facilitation networks, with conventional forces, which were attacking network sanctuaries and freedom of movement and reestablishing the ISF presence along the border. The task force had established its own country-wide intelligence collection and operational network to go after al Qaeda in Iraq. Its efforts were focused on AQI leadership, and it conducted several operations a night across Iraq in pursuit of al Qaeda targets. It coordinated its efforts with local commanders in whose areas it operated. This coordination improved over time as the conventional and special operations forces became more comfortable working together. The targeting process developed by the task force proved very effective for hunting down individual terrorists, so, with its assistance, we began to develop "fusion centers" in each of the U.S. divisions in 2005. These fusion centers enabled the coalition divisions to access intelligence from all available national sources and, because they were directly connected to forces that could rapidly act on the intelligence, to attack high value targets in their areas of operations independently. These centers greatly increased our ability to attack al Qaeda and insurgent leadership and were instrumental in our long-term success.

By the end of the year, we had significantly disrupted the flow of foreign fighters and suicide bombers into Iraq and seen the number of suicide attacks cut in half between June and November to the point that the constitutional referendum and December elections were held with limited interference. (There were 90 and 80 attacks, respectively, on referendum and election days compared with 299 on the day of the January 2005 elections.) The operations also loosened the al Qaeda stranglehold on the Sunni populations of those regions, allowing them to participate more fully in the referendum and election processes—something that would help us later.

We continued to monitor the implementation of our transition team and partnership programs and continued to evaluate ourselves on how we were applying counterinsurgency doctrine. When we implemented the transition team concept in the spring, I directed a special assessment of the concept for September. By September, we had embedded 174 transition teams into Iraqi military and police units, and seen increased performance in those units as a result. We had also developed and instituted a transition readiness assessment (TRA) to measure and quantify the readiness of Iraqi units to assume security responsibility.* The assessment found that the transition teams, augmented by the partnership program between Iraqi and coalition forces, had "made a significant difference in our efforts to rebuild and professionalize the Iraqi Army." It also found that the Iraqi security ministries (MOI and MOD) showed "limited progress toward self-reliance" and that the Iraqi Police Service "lags in development." We had a long way to go, but I felt fairly comfortable that we had a credible system to measure progress and the presence in the Iraqi units to verify it. The assessment also found that our presence was positively received by Iraqi units and that it served as a deterrent to detainee abuse and violence against civilians.[9]

We used this assessment to press the Ambassador to move the responsibility for developing the ministries of interior and defense from the Embassy to MNF-I. I believe that this move made a significant difference in our ability to increase the capacity of the ministries. We quickly began a serious effort to develop key ministerial functions in both ministries—planning, programming, budgeting, manning, equipping, and sustaining—and to establish internal accountability. We also used it to increase our focus on the development of the police. We calculated that the police were about a year behind the army in terms of development. We believed that we could not credibly handoff security until the local police were capable of maintaining domestic order and denying terrorist activity, so we began to develop a plan to

---

* By this measure, 1 Iraqi division headquarters, 5 brigades, and 7 battalions had achieved the second highest transition readiness assessment—able to conduct counterinsurgency operations with limited coalition support—by that June.

accelerate police development in 2006. Police development is much more difficult than army development, especially for soldiers. This is an area that needs continuous work across government to craft the means to more rapidly build police capacity. With the development of the other rule-of-law institutions (for example, judicial and prison systems), it was the long pole in our development tent.

At the same time, we completed a survey of how we were applying counterinsurgency doctrine across the force. As I observed the new forces coming into Iraq, it seemed that our execution was uneven at best, as we and the Services wrestled with ingraining a new form of warfare into our conventional forces. I sent a team across MNF-I for 15 days that summer to take a look. The team concluded that there was a general understanding of counterinsurgency doctrine, but that its application was in fact uneven and very dependent on the individual commander's grasp of the doctrine and how to apply it in Iraq. It also concluded that we were still being forced to apply peacetime practices in a wartime environment. As a result, things got even more complicated the further down the chain of command we went as each level added its own restrictions. It was a good and thorough report that I forwarded to the Service chiefs for their use in training deploying forces and to my staff and subordinate commanders to begin implementing the recommendations. One of the principal recommendations was to establish a COIN Academy to teach the nuances of applying counterinsurgency doctrine in Iraq to incoming commanders in order to ensure more commanders started at the same level. We established the academy and conducted the first course that November. I addressed every class, and felt that this academy not only substantially improved our execution in Iraq, but also formed a basis on which the Army and Marine COIN doctrines could be updated. It was a significant factor in changing the conventional mindset of U.S. Servicemembers.

In the meantime, the Iraqis, with coalition support, conducted the second of the UNSCR-prescribed national polls, a referendum on the Iraqi-drafted constitution. Over 15 million citizens, nearly 64 percent of those eligible, registered to vote, and 10 million of these citizens voted, an increase of 1.5 million from the January election. On referendum day, over 6,000 voting sites were open, more than a 20 percent increase from the previous election, with major increases in the Sunni areas. Violence levels were a third of the previous January, with only one-quarter the casualties. In a radio address to the American public, President Bush stated, "By casting their ballots, the Iraqi people [dealt] a severe blow to the terrorists and [sent] a clear message to the world: *Iraqis* will decide the future of their country through peaceful elections, not violent *insurgency*. And by their courageous example, they are charting a new course for the entire Middle East."[10] As I reported to both President Bush and Secretary Rumsfeld, the referendum's success was the result of a well-executed civil-military plan with extensive preparatory operations.

Though 78.6 percent of Iraqi voters cast ballots in favor of the constitution, Sunni voters largely voted against it. The Sunni representatives had been marginalized during the drafting process, and the constitution did not emerge as the national compact that we had hoped for. In fact, it was only the last-minute efforts of the Ambassador to elicit promises to address Sunni concerns and amend the constitution after the vote that made Sunni participation even possible. What could have been a major step forward was not. So while we were pleased with voter participation and ISF performance, we were apprehensive about the long-term impacts of the new constitution.

COMPLETING THE UN TIMELINE

Planning for the Future. We were also generally pleased with the implementation of the transition team concept and the positive impact it was having on the growth of the Iraqi security forces. We had greatly expanded our insights into their capabilities from the ministry to the battalion levels. What remained was to translate the increase in Iraqi capability into their assumption of security responsibility across Iraq as envisioned by UNSCR 1546, and then to link that to the gradual reduction of coalition presence.

The transition concept gained explicit Iraqi acknowledgment in July, when Prime Minister al-Jafari, after a visit to Washington, announced the formation of a Joint Committee to Transfer Security Responsibility to establish the conditions for gradual transition of security responsibility to the Iraqi government. The committee consisted of seven U.S., UK, and Iraqi senior-level principals (the United States was represented by Ambassador Khalilzad and me), who oversaw the efforts of a joint working group that developed the specific conditions under which security authority would transfer. It was also intended that the committee would monitor the implementation of the process and make recommendations to the prime minister regarding transfers over time. The committee was chaired by the prime minister, and after several months of work it produced a document that laid out the conditions for the assumption of security responsibility by the Iraqis in a manner that would maintain security against the terrorists during and after the transitions.

As part of this process, we established criteria that assessed the ability of the security forces—army and police—to work together with the provincial leadership to maintain order and defeat terrorist and insurgent threats in and around the province. The evaluation process involved an assessment of the threat; an assessment of police capability in the province (all had to be rated at least "TRA 2,"* that is, able to conduct COIN operations with limited coalition support); an assessment of the military capability in the province (all had to be TRA 2 and able to coordinate operations with the police); and an assessment of the provincial leadership's ability to coordinate security efforts. We worked hard to find the right conditions that would serve as a forcing function for the Iraqis to increase their capabilities and yet still be attainable. We intended for all transfers to be conditions-based and therefore did not set a timeline for the transitions. There were no provinces ready for transfer in 2005, and we did not get our first chance to implement the process until the summer of 2006 when Muthanna Province in southern Iraq became the first Iraqi province to assume responsibility for its security.

In September, Secretary Rumsfeld requested that I provide Stephen Hadley, the National Security Advisor, with an update of our campaign. There was apparently some uncertainty with what we were doing to accomplish our national objectives in Iraq. Although puzzled by this, as we were providing weekly updates by video teleconference to the National Security Council and the President, I provided Mr. Hadley with an overview of our campaign plan from July 2004 through September 2005. I emphasized how the campaign had evolved over the past year as we adapted to changes in the threat and the environment, how our plan had changed with the implementation of the transition teams, and how the military operations we were conducting had kept the pressure on the insurgents and terrorists while

---

* *TRA* is a monthly report prepared by both the Iraqi commanders and transition team leaders that documented and rated ISF unit ability to conduct independent COIN operations. TRA 2 was the second-highest rating for a unit.

we worked to complete the implementation of the UN timeline and grow the ISF.[11] I did not receive any new direction as a result of the discussions, and at the end of November, the National Security Council (NSC) issued the National Strategy for Victory in Iraq (NSVI). Its endstate was almost the same as the one that Ambassador Negroponte and I had crafted 18 months prior: "A new Iraq with a constitutional, representative government that respects civil rights and has security forces sufficient to maintain domestic order and keep Iraq from becoming a safe haven for terrorists." The description of "victory" in the strategy paralleled our own lines of operation with political, security, and economic tracks. On the political track, the NSVI prescribed forging a broadly supported national compact by isolating Iraqi elements that could not be won over to the political process, engaging those outside the political process, and building effective national institutions to protect all Iraqis. The economic track prescribed setting the foundation for a self-sustaining economy by restoring Iraq's infrastructure, reforming Iraq's economy, and building the capacity of Iraqi institutions to maintain infrastructure. The security track stated that our strategy was to "clear" areas of enemy control by remaining on the offensive, "hold" areas freed from enemy influence, and "build" ISF and the capacity of local institutions to deliver services.[12] The strategy had codified our approach.

Shortly after receiving the NSVI, the Ambassador and I issued a new joint mission statement, "Building Success—Completing the Transition." This statement was the output of a joint planning effort that had been ongoing since the summer to devise a strategy and plan to guide coalition and Embassy operations following the election and seating of the new Iraqi government. Significantly, we took a longer view than the first campaign plan and tied our guidance to the 4-year term of the soon-to-be-elected government.

We set our objectives for the mission during this period as:

- defeating the terrorists and neutralizing the insurgency
- transitioning Iraq to security self-reliance
- helping Iraqis forge a national compact for democratic government
- helping Iraqis build government capacity and provide essential services
- helping Iraqis strengthen their economy
- helping Iraqis strengthen the rule of law and promote civil rights
- increasing international support for Iraq
- strengthening public understanding of coalition efforts and public isolation of the insurgents.

We further directed annual goals to guide our progress. Our planners took this directive and began to turn it into a campaign plan to guide our efforts for the next 4 years.[13]

In December, we held our semi-annual Campaign Progress Review,[14] which, for the first time, included the full participation of the Embassy staff. The review concluded that there were "clear grounds for optimism" as we had successfully completed the 18-month UN timeline, thwarting terrorist and insurgent efforts to derail the political process, and had developed the ISF to the point that they principally provided security for their elections. This was a significant accomplishment, and one that I was not sure we would accomplish 18 months before. We had also made great progress in growing the ISF, particularly on the army side. By the end of 2005, 80 percent of Iraqi battalions were fighting the insurgency with us. Moreover, 1 of the 10 divisions, 4 of the 36 brigades, and 33 of the 112 battalions were able to operate with limited coalition support—a significant increase in just a year. But we were a long way from being done. The police lagged the development of the army significantly and the institutional capabilities of both the interior and defense ministries were still in their nascent stages. It was clear that we needed a major effort in both these areas to accelerate their development.

Perhaps the most troubling trend at the end of 2005 was the increase we were seeing in sectarian violence. On 14 September, Abu Musab al-Zarqawi issued a public declaration of war against "Shia infidels" while continuing his campaign of suicide attacks against the Shia population. Although our operations in the west had greatly reduced the number of suicide attacks, there were still enough attacks to drive sectarian tensions. We were also seeing small-scale actions by Shia militia against the Sunni population. The trend was troubling enough that we established an intelligence working group that September to monitor it. These tensions were compounded when, in November, coalition forces discovered an MOI detention facility in which the detainees, primarily Sunni, were mistreated and in some cases tortured. Sunni leaders clamored for action against the interior minister, but the leadership of the transitional government took none. It was becoming clearer to us that just completing the UN political process was not going to be enough to bring the country together. We came to believe that a program of national reconciliation would be essential if Iraq was to move forward.

The year ended with the elections of December 15 where 11.8 million Iraqis (76 percent of registered voters) elected a government based on the constitution that they had approved in October. While the government had yet to be formed, we had completed the mission that we had set out to accomplish 18 months before.

It had been a tough 18 months. I was pleased with what we had accomplished, yet I recognized how far we still had to go to defeat the insurgency and credibly hand over the security mission to the Iraqis. It remained to be seen whether our key assumption would hold true—that the completion of the UN political timeline would yield an Iraqi government that was perceived as legitimate and representative of all Iraqis. The constitution had not been the national compact we had hoped for because the Sunni representatives were largely excluded from the drafting process. And while there was talk of forming a "government of national unity," it remained to be seen if the major parties could set aside their sectarian fears and prejudices for this to happen.

We closed out 2005 feeling that we had accomplished a great deal, but that we still had much more to do—and we would do it with a new, as yet undetermined, Iraqi government, a new set of coalition forces, and without a political timeline to drive Iraqi action. I worked to dampen Washington's "optimism" over our substantial accomplishments as we worked to bridge what we knew would be a difficult period of governmental transition.

Prime Minister Ibrahim al-Jafari addresses city leadership of Fallujah January 2006 as Ambassador Zalmay Khalilzad and General Casey look on

4. GOVERNMENT TRANSITION AND THE RISE OF SECTARIAN VIOLENCE (JANUARY–JUNE 2006)

We entered 2006 knowing that it was going to be a year of political transition as the Iraqis formed and seated a permanent government and this new government began to govern. We remained hopeful that we could continue to make progress during this period; however, based on our previous experience with Iraqi government transitions, we knew it was going to be hard work. That was the message that I carried to Washington shortly after the first of the year. I spent Christmas in Iraq visiting and congratulating soldiers on their accomplishments in 2005, and then departed for consultations in Washington.

I reported that, in general, we were pleased with the accomplishments of the last 18 months, but the elections had not yet produced the representative government that was key to long-term success. Ambassador Khalilzad felt that the elections had a polarizing effect in the country. The Sunni population felt disenfranchised as a result of the constitutional referendum and the election outcome, and their perception of increased Iranian influence on what was, almost certainly, to be a government formed by largely Shia political parties compounded these feelings—and fed the insurgency. I cautioned not to expect any immediate positive impacts on the security situation and warned that political wrangling over amending the constitution and the provincial elections could even push things in a more negative direction.

I saw our main challenge in 2006 as getting the new government, particularly the security ministries, onboard and governing as rapidly as possible. This would allow us to take advantage of its 4-year tenure and build the ministerial capacity that Iraq would need for the long haul. On the security side, I was concerned with the increase in sectarian violence that we had been tracking and its potential negative impact on government and security force development, and with the lagging development of the police—which we knew was the key to our long-term success. On the positive side, we had good success in disrupting al Qaeda in Iraq over the past year and were beginning to see a rift between them and the Sunni population that we hoped to exploit. We also continued to make good progress with the Iraqi army.

I concluded my discussions in Washington with a slide entitled "Bad Things That Could Happen"[1] to remind everyone that, as much as we had accomplished in the past 18 months, we were in a time of political drift in a violent, divided society and bad things

GOVERNMENT TRANSITION

could happen that could affect the direction of the mission. I thought it best to temper expectations because there was just too much uncertainty in this government formation period. Looking back, several of the things on the list did happen, and did affect the direction of the mission.

The other major element of discussion in Washington was the implementation of the first "off-ramp" plans that we had announced following the completion of the elections in December.* I had recommended, and the Secretary of Defense had approved, not replacing 2 of the 17 U.S. Brigade Combat Teams that were programmed to redeploy from Iraq that summer, effectively reducing the number of U.S. combat brigades from 17 to 15. This was a difficult decision, as we were very cognizant of the uncertainty and potential turbulence of the post-election period, but we felt that the improvement, and projected improvement, of the ISF would more than mitigate the tactical risk.† We also believed that the potential strategic benefits far outweighed the tactical risks. The image of U.S. forces departing Iraq would demonstrate to the ISF and the Iraqi people that the United States was indeed serious about following the UN mandate to return security responsibility to the Iraqis as they became more capable, and about ultimately departing Iraq. We decided to keep one of the brigades in Kuwait as an in-theater reserve just in case we had miscalculated. The discussions that January revolved around developing a recurring process to periodically review the situation and make recommendations on the continuing reduction of U.S. forces.

---

* *Off-ramp* was the term used to describe the removal of troops from theater without replacements, essentially reducing U.S. forces.
† At the time of the decision, there were some 216,000 trained and equipped ISF.

I recommended a quarterly, conditions-based process, and the first assessment was set for March.

Government Transition

While I was back in the United States, I was asked to extend for another year until the spring of 2007. I had wanted to see the establishment of the new government through, so I agreed, with the caveat that this would be the last extension. I returned to Iraq intent on maintaining positive momentum during the government transition period in any way that I could. In October, we had established our Bridging Strategy—al Qaeda out, Sunni in, and ISF in the lead—to guide our actions during the transition period. It was also intended to provide focus for the new MNC-I during the early months of 2006.

Al Qaeda Out. Our operations in the western Euphrates valley and Tal Afar–Mosul corridor had significantly disrupted al Qaeda facilitation networks in those areas, and our special operations task force had kept strong pressure on al Qaeda leadership. Interestingly, al Qaeda's brutality and desire to impose sharia law in areas they controlled began to wear on the Iraqi population, particularly in Anbar Province. That winter we began efforts to consolidate our gains in Anbar. On January 15, the Ambassador and I took Prime Minister al-Jafari to Ramadi and Fallujah. The leaders there told us that they were not happy with the constitution and wanted an end to the occupation, but that they were against terror. They wanted an army division recruited strictly from Anbar Province, money for reconstruction, and the release of all Anbari prisoners from confinement. The prime minister agreed to $75 million for reconstruction and promised to send the ministers of defense and interior to discuss

GOVERNMENT TRANSITION

the security force proposals. The agreement was finalized between the al-Jafari government and the governor of Anbar Province over the next 2 months and was implemented over the course of 2006.

The ISF plan built on a small success that we had with the Albu Mahal tribe just prior to the December elections in forming the "Desert Protectors" to assist in securing the border region of Anbar. While the leaders of Anbar did not get their Anbar division—we were concerned with creating a strictly Sunni formation—they were allowed to bring almost 16,000 police and army recruits into the ISF. Each was vetted and vouched for by his tribal sheik. They generally performed well in Anbar, particularly against al Qaeda. It took constant pressure on the government to distribute the promised reconstruction dollars, but they finally did, and small prisoner releases were conducted over the course of the year. This effort marked a major turning point not only against al Qaeda, but with the population of Anbar Province.

Sunni In. As a result of Ambassador Khalilzad's hard work on the constitution, we had a major breakthrough with Sunni political leaders just prior to the elections that we hoped to exploit in the new year. The Sunni political leaders came to the Ambassador a few days before the election and stated that they wanted to tell their people to vote on December 15, but they were afraid the people would get caught up in ongoing operations. They wanted me to announce the suspension of operations for the elections in exchange for their support in turning out the Sunni vote. I confirmed with my divisional commanders that their major operations were completed, and they were. We had planned all along for our pre-election operations to end a few days before the election. I told the Sunni leaders that we

STRATEGIC REFLECTIONS

would not conduct any major operations in the last days before the elections but that we would continue routine force protection and high-value target operations. That was good enough for them, and, true to their word, they went on Iraqi television and told Sunni voters to vote.

The Sunni leaders told us that January they had concluded that their common enemies were al Qaeda and Iran—not the United States. They felt that we were leaving, and they needed our help to get Iraq back from al Qaeda and Iran. The election collaboration was a small step and the beginning of a series of confidence-building measures between the MNF-I and Sunni political leadership that led to serious, but ultimately unproductive, discussions about ceasefires and reconciliation over the course of 2006.

To continue this dialogue, we established a cell, under a U.S. two-star general, to better coordinate engagement activities with political and insurgent leaders among the different agencies of the mission. At this time, the U.S. Embassy, UK embassy, CIA station, and MNF-I were all getting feelers from different leaders and groups claiming to have influence over the insurgency. Each needed to be vetted, evaluated, and acted on in a coordinated way. While there were some hopeful signals, several promising leads failed because of the inability of the interlocutors to deliver on their promises. It soon became clear to us that this was an early but essential part of the ultimate reconciliation process that the Iraqis would have to go through to conclude the insurgency. In the end, it was the beginning of an important dialogue that we continued. It marked a significant shift in how the Sunni population, particularly in Anbar Province, saw the MNF-I and U.S. presence.

ISF in the Lead. The growth of ISF was just about on schedule at this point (mid-January 2006) with over 225,000 trained and equipped army and police forces of the 325,000 that we were building with the Iraqis. We believed at that time that it would take us another year to complete the training and equipping of the ISF and that the police would take a concerted effort just to make that. More significant was the increase in the number of these forces that were actively participating in counterinsurgency operations either independently or with coalition support—100 Iraqi army battalions were actively in the fight, a number that had increased by one-third in just 6 months as the transition teams and partnership programs continued to demonstrate their worth.

The police and security ministries continued to lag the army in development. We began a focused effort to improve police capacity and planned for the transition in ministry leadership that would come with the establishment of the new government. The deficiencies in the police were well known. Our trainers were reporting a lack of trained police, significant deficiencies in low-level leadership, and sectarian bias at the highest levels of the Ministry of the Interior. Our effort to improve police capacity became known as the "Year of the Police." We introduced specific programs not only to increase the numbers of Iraqi police, but also, more importantly, to improve their professionalism and quality of their training. Our 2004 COIN study had driven home the key role to be played by the police, so we needed to refocus our efforts to organize, equip, train, and field a professional police force that embodied national unity. The ambitious goals we set for ourselves—an MOI capable of leading border security efforts by June 2006 and assuming the lead in the counterinsurgency effort by the end of 2006—were indicative of the importance we placed on the role of the police in our long-term plans. The most difficult challenge proved to be eliminating sectarian bias from the ministry. This was essential because of the debilitating impact it had on the even-handed enforcement of Iraqi law and the confidence of the Iraqi people in their police forces. It would significantly delay our efforts with the police.

The other key aspect of sustaining our momentum was the smooth transition of MNC-I in January. Our preparatory training was very good, but it was simply not possible to replace a year's experience on the ground overnight, particularly when we were asking our conventional forces to do things that they had not done for decades. While I saw continuous improvement in the preparation of new forces over time, and our COIN Academy helped strengthen the transition process, transitions were always periods of friction and turbulence.

The new MNC-I commander, LTG Pete Chiarelli, was returning to Iraq after only a year away, having commanded the 1st Cavalry Division in Baghdad from April 2004 to March 2005. In our initial meeting, I emphasized the importance of maintaining momentum during the government transition process and asked him to give me his preliminary assessment of the situation in 30 days. In that assessment, he commented on the significant progress of the Iraqi army, particularly at the tactical level, and the encouraging improvement in Anbar Province. He expressed concerns about the lack of development of the police and in the capacity of the governmental ministries and with militias challenging the authority of the ISF.

He also commented on the substantial improvements that had been made in the counter-IED (improvised explosive device) effort

GOVERNMENT TRANSITION

after he left. With the establishment of a counter-IED task force in 2005, MNC-I focused on attacking IED networks as a system by bringing technology to bear for tracking IED emplacers, exploitation of sites, and jamming detonators. This approach and an influx of hardened vehicles allowed the corps to increase the number of IEDs found and cleared by almost 15 percent and to reduce significantly the effectiveness of IED attacks.* IEDs would continue to produce the largest number of casualties among coalition forces and would require continuous adjustment and adaptation by coalition leaders and soldiers.

As we headed into the third month of government formation in mid-February, we continued to press ahead with our Bridging Strategy, settling in the new corps, and dealing with increasing signs of sectarian violence, while the Iraqis moved frustratingly slowly toward forming a government.

The Samarra Bombing and Its Aftermath

On February 22, the al-Askari Mosque, a sacred Shia holy site in Samarra, a Sunni city north of Baghdad, was destroyed by a bomb, unleashing a spate of sectarian violence against the Sunni population in Baghdad and the surrounding areas. Shia militia attacked Sunni neighborhoods and mosques, causing widespread panic in the capital. Coalition forces reacted quickly to staunch the violence while the Ambassador and I attempted to mobilize the Transitional Government to act. Prime Minister al-Jafari was slow to impose an evening

---

* As a result of these efforts, the average casualties per improvised explosive device detonation went from 0.95 casualties per detonation in the June–December 2003 timeframe to 0.35 casualties per detonation in the January–February 2006 timeframe, a significant improvement.

curfew that would have facilitated our operations. While the government sought publicly to portray the attack as an attack against the Iraqi people by the enemies of Iraq, it seemed clear that the attack was viewed by the Iraqi leadership as a direct attack on Iraq's Shia population, and this seemed to slow their initial reactions—which was not lost on Sunni leaders. While the government eventually imposed the necessary curfews, it continued to resist a ban on the carrying of weapons by other than ISF and would not implement policies to prohibit nongovernmental militias.

We saw the sectarian violence unleashed by the attack as having the potential to threaten our ability to make progress in 2006 by exacerbating the fraying tensions between sectarian groups and making the formation of a government of national unity more difficult—further exacerbating tensions and extending the government formation process. I was very concerned about this escalating into a major sectarian conflict and causing a fracture of the ISF along sectarian lines—two things that could fundamentally change the course of our mission in Iraq. I issued orders to isolate and stabilize the situation in Baghdad, focusing on ethnically mixed areas, and to work on preventing attacks that could further inflame sectarian tensions.[2] Meanwhile, the Ambassador and I worked with the Transitional Government to get the needed political support to frame a security operation to secure Baghdad. The next few days were difficult as we guided the now lame-duck government through a major crisis while our forces worked to restore calm to the capital.

After a few days, we persuaded the prime minister to begin a concerted effort to restore order to the capital and surrounding areas where the vast majority of the sectarian violence was taking place. The intent of the operation, called *Scales of Justice*, was to stop the sectarian attacks and provide sufficient security in the capital region so that the new permanent Iraqi government could be formed and seated. At that time, we still held out hope that the formation process would produce a government of national unity that would be perceived as representative of all Iraqis, or at least as more representative than the Transitional Government. We pushed eight more Iraqi and coalition battalions into the key districts of Baghdad, massing more than 50,000 coalition and Iraqi security forces into Baghdad and the surrounding areas. The operation stabilized the situation and bought us the time to complete the government formation process. However, after initial military success, *Scales of Justice* slowed, but did not halt sectarian violence in Baghdad, leaving a difficult situation for the new government.

The operation was hampered by the reluctance of the Transitional Government to impose weapons and militia bans and to stem sectarian influence within MOI forces. This reluctance had a very negative impact on the confidence of the Iraqi people, particularly the Sunni population, in the ability of the police to evenhandedly enforce the law. We suspected that some police formations and political militias were actively conducting sectarian killings with at least the tacit approval of political leaders. In all, coalition forces, and a good number of their Iraqi counterparts, did a remarkable job of stabilizing a very difficult situation, but their increased exposure resulted in an increase in coalition and ISF casualties.

Assessing the Impact. Once we got the military response moving, I turned my attention to determining the impact of the Samarra bombing and its aftermath on our long-term plans. It seemed clear that the sectarian tensions we had seen emerging since the summer of 2005 with the seating of the Transitional Government and the

September declaration of jihad against the Shia population by al-Zarqawi were more than just tensions and needed to be addressed more broadly.

From my personal interaction with Iraqi leaders, I saw a great fear of "Ba'athist return" among Shia leaders: a fear that the Ba'athists would overthrow the duly elected Shia government and return to power, subjugating the Shia population as they had been for more than 30 years under Saddam Hussein. I also saw a great fear among Sunni leaders that Iranian influence would ensure the continued subjugation of the Sunni population. While we might have seen these views as caricatures, they defined the views of Iraq's leaders, who would ultimately have to resolve these sectarian tensions. These views were not new. What was new, however, was how the scope of the sectarian violence made them more real to the Iraqis and made trust between Iraq's leaders more difficult to attain.

We began a review of our plans and strategy, asking ourselves the fundamental question: "Does the advent of significant sectarian violence in this transition period require a change in our strategy for Iraq, and, if so, what should we do differently?" Looking at the situation, we came to the conclusion that the post-Samarra violence may have been an indication of a significant change in the nature of the Iraqi conflict. Taken in the context of the conclusion of the UN political timeline, the initial departure of coalition forces, and the ongoing efforts to form a 4-year Iraqi government, the sectarian violence seemed to indicate that the main conflict in Iraq was moving away from an insurgency against the coalition to a struggle for the division of political and economic power among Iraqis. This would be a significant change.

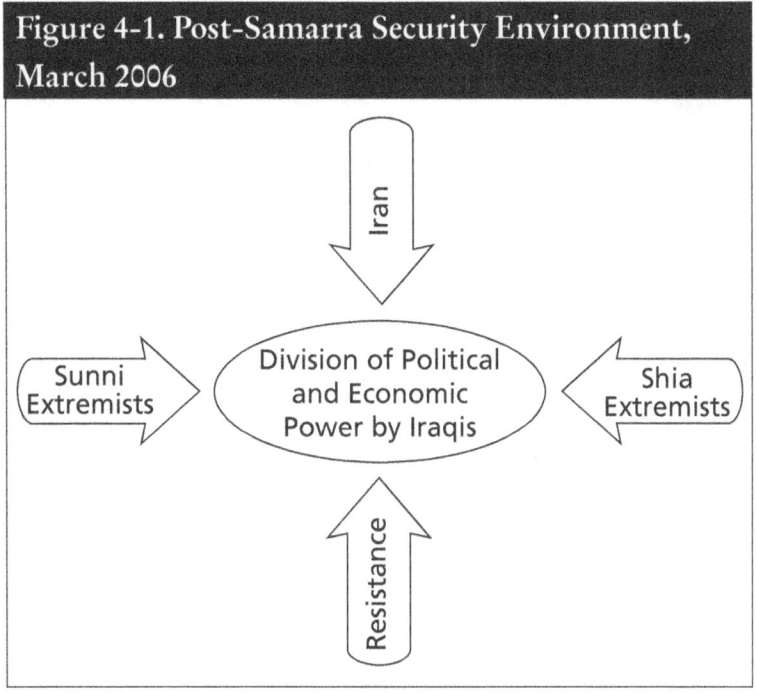

Figure 4-1. Post-Samarra Security Environment, March 2006

As I thought my way through this, I saw four major groups influencing the security situation to affect the outcome of the conflict: Sunni extremists, Shia extremists, the Sunni resistance (the insurgency), and Iran. I saw the main antagonists as the Sunni and Shia extremists, each heavily influenced by fear of the other and each motivated by fear of exclusion and retribution. The Sunni resistance, which was still active against the coalition in parts of the country, was losing relevance and influence. Finally, the Iranians were actively supporting Shia extremist groups and using political and economic influence to shape an outcome that would ensure that the new Iraq would be a benign neighbor. I graphically portrayed this as shown in figure 4-1. It was a far more complex environment than we had previously dealt with.

I discussed this view with my commanders and the Ambassador. They generally agreed that the nature of the conflict had changed. The commanders noted that they were dealing with different parts of the conflict in their different parts of the country, and that the main struggle was going on in Baghdad and the surrounding provinces. I developed the graphic below (figure 4-2) to communicate to my commanders the actions we needed to take. We needed to block the ability of the Sunni and Shia extremists and Iran to influence the division of political and economic power, while working to bring the Sunni resistance into the political process. To do this, on the military side we would have to work with the Iraqis to provide an environment secure enough for political and economic development to continue and in which the population felt protected and civil war was averted.

We also did some work to explore various courses of action that we could take if the sectarian tensions spilled over into full-scale civil war. There was a lot of discussion with Washington in March of 2006 about whether we were in a civil war, how we would know it if we were, what civil war would look like, and most importantly, what we would do if we were. I felt strongly that the sectarian violence spawned by the Samarra bombing had not reached the level of civil war primarily because the violence was confined to such a small portion of the country, the ISF remained a national force, and there was still support for a political process. That said, it was clear that the extremists were attempting to push the country toward civil war as a means of achieving their political objectives, and it was in our interest to prevent that from happening. In response to a request from Washington for

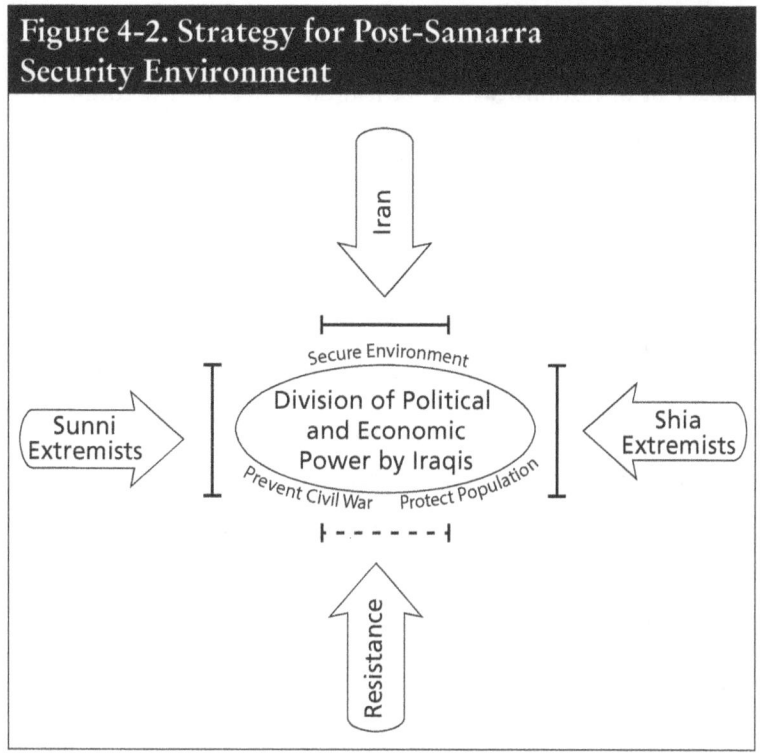

Figure 4-2. Strategy for Post-Samarra Security Environment

my thoughts on what to do if the violence did degenerate into civil war, I stated that we could leave, return to our bases and wait it out, intervene, or pick a side, but that doing anything but reinforcing and stopping the violence—intervening—would acknowledge strategic defeat.

We also got a number of questions during that time from Washington about how U.S. forces were operating and how we might shift our weight away from activities that caused casualties. We responded with a study that looked at whether having increased coalition presence in the streets actually improved security or if our higher visibility posture simply invited attacks and reduced the pressure on the Iraqis to step forward.

Our study showed what we intuitively knew: that increasing the number of operations, which we had done in response to the sectarian violence, did not necessarily translate into higher casualties. Rather, maintaining an offensive posture kept the enemy off balance and disrupted his ability to respond to our actions. It also allowed us to maintain necessary contact with the population and to conduct the necessary patrolling for force protection and intelligence operations that were essential to successful COIN operations. We also found that coalition casualties decreased when security responsibilities were transferred to the Iraqi army. The majority of coalition and ISF casualties were confined to just 3 of the 18 provinces, namely Baghdad, Anbar, and Ninewah. Civilian casualties were mainly limited to just two provinces: Baghdad and Diyala.

Charting a Way Forward. While we dealt with the Samarra adjustments, we moved ahead with the development of our overall campaign plan made necessary by the completion of the UNSCR political timeline. The outcome was a comprehensive effort with the Embassy designed to guide the mission over the next 3-plus years—the tenure of the new, still-to-be seated government.

We defined these next years as "the decisive phase to bring security and stability to Iraq" and laid out three phases: Phase one—stabilization (2006 to early 2007); Phase two—restoration of civil authority (early 2007 to early 2008); and Phase three—support to self-reliance (early 2008 to 2009). Using this strategic framework (see figure 4-3), we laid out specific objectives and tasks in each phase that were aligned along the five integrated lines of operation, much as we had outlined in our earlier campaign plan—security, governance, economic development, communications, and transition (added in the new plan).

GOVERNMENT TRANSITION

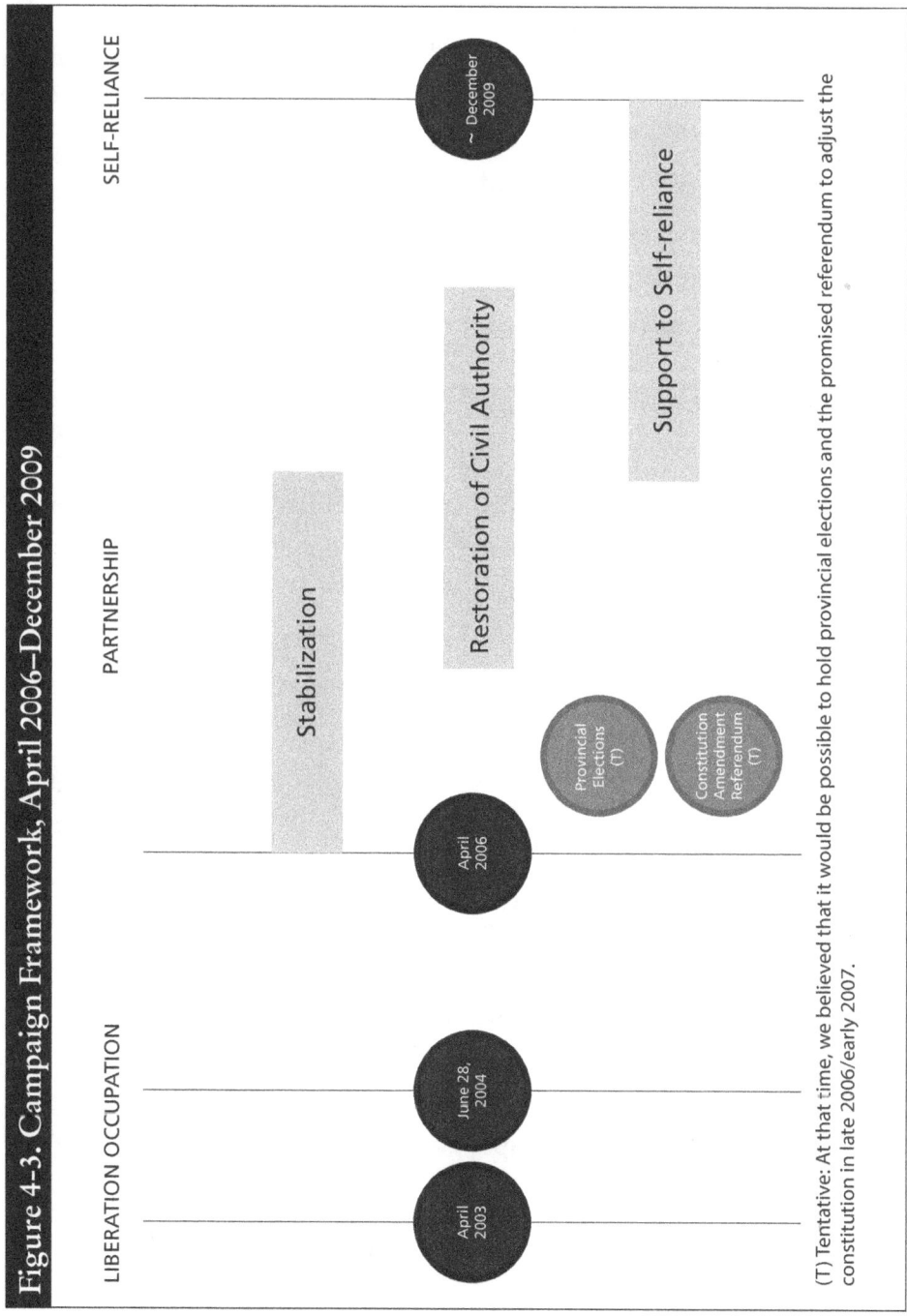

Figure 4-3. Campaign Framework, April 2006–December 2009

(T) Tentative: At that time, we believed that it would be possible to hold provincial elections and the promised referendum to adjust the constitution in late 2006/early 2007.

We also looked at "wild cards" that might impact the projections we made for each phase, established specific effects and metrics for each phase, and identified the coalition or U.S. Embassy agency responsible to the Ambassador and me for accomplishing them.[3]

Most importantly, we included the strategic concept developed in the aftermath of the Samarra bombing into the campaign plan. In it we adjusted our mission statement to reflect the new situation:

> *The U.S. Mission and Coalition Forces will, in partnership with the Government of Iraq, contribute to an environment where Iraqis can develop representative and effective institutions capable of meeting the needs of the Iraqi people, creating the conditions for the Rule of Law, defeating the terrorists and irreconcilable insurgents, bringing other insurgents into the political process, reducing sectarian tensions and denying Iraq as a safe haven for terror.*[4]

The mission and the campaign plan reflected our central belief that the conflict was about the division of political and economic power among Iraqis. We believed that enduring strategic success could only be achieved by Iraqi political and military leaders working together to resolve Iraq's substantial problems. It was our job to work with them to help them do that.

Building New Partnerships

There was a break in the political stalemate in late April when a relatively unknown Shia politician, Nuri al-Maliki, was appointed prime minister. While it would take another month before he and his cabinet were sworn in, and 2 weeks after that before agreement would be reached on his key security ministers, it was a start.

During the month between the appointment of Prime Minister-designate al-Maliki and his formal inauguration, the Ambassador and I met with him frequently to update him on pressing issues and to get his views on the way ahead. We recognized from the outset that we had to build our relationships with him, just as we had with his predecessors, if we were going to sustain a partnership to deal with the very difficult issues facing the country. What I did not foresee, however, was how much of my personal time and energy would be consumed in building and sustaining my relationship with the new prime minister in the coming months.

In our first session, I addressed the security situation and development of the ISF. I told him that the primary threat to Iraq was terrorists and militia fomenting civil war and that Baghdad and Anbar provinces were the greatest security challenges. I recommended developing and implementing a plan to secure Baghdad as the first priority. I told him that the ISF were making good progress in building security capacity, but that we needed a major effort to restore public confidence in the police, particularly among the Sunni population, because of their perceived, and sometimes demonstrated, sectarian bias. I also told him that we were 18 to 24 months away from having the ISF to the point where they could operate without substantial coalition support. I encouraged him to build a strong, representative security team and to address the militia issue as matters of priority. He stated that he thought the 18 to 24 month timeline for the security forces seemed like a long time, and shared his view that the ISF were poorly trained and equipped, and in some cases infiltrated. I knew I would have to bolster the prime minister's confidence in his ISF.

During another of our initial meetings, the Ambassador and I drew on our experiences with the previous transition and offered some thoughts for the first 100 days of the new government. We suggested a program based upon three tenets—Unity, Security, and Prosperity—and provided him with some recommendations to generate political momentum in these areas during his first days in office. We believed that the new government had to be perceived as representative of all Iraqis and seen as taking positive steps to reconcile the concerns of the ethnic and sectarian groups (Unity) as a complement to security efforts to halt the violence (Security). Once security was established, economic development could continue (Prosperity). We had numerous other sessions with Prime Minister al-Maliki in this interim period to ensure this wartime transition of power went as smoothly as possible.

Prime Minister al-Maliki and the majority of his cabinet were sworn in on May 20. The key security ministers could not be agreed upon in time for the parliamentary session, so the selection process for the security ministers continued into June. In our first meeting with the newly inaugurated prime minister, he agreed to the Unity-Security-Prosperity construct that the Ambassador and I had laid out earlier. He stated that he wanted our help to formulate a short-term plan to improve current conditions, and a longer term plan to resolve the more intractable issues facing Iraq. He asked to meet in a few days to discuss a plan that would lead to a "dramatic" improvement in the security situation in Baghdad. He also expressed concern about the situation in Basra where competition for wealth among the

Shia population and increasing Iranian influence were making for a difficult security situation.

With so many pressing issues, the Ambassador and I worked to get only the most important issues in front of the prime minister to avoid overwhelming him in his first days. On the security side, I needed to help him assume his role of commander in chief of Iraqi forces, work with him to develop and take ownership of plans to secure Baghdad, familiarize him with the capabilities of his security forces, and, further down the road, familiarize him and his government with the plans to transition security responsibility to capable ISF. There were also two time-sensitive issues—the need to strike a Sadrist headquarters near Sadr City involved in kidnapping and murder, and a coming meeting in Baghdad with the Iranian foreign minister—that we needed him to focus on.

Earlier that month, one of our unmanned aerial vehicles had filmed a kidnapping and murder in Baghdad from start to finish. A man walking across a bridge was forced into a car and taken to a large walled complex that we knew to be a Sadrist headquarters. After his captors went inside for a period, they came out and drove the man to the outskirts of Baghdad where they shot him. The Ambassador and I took the videotape to the prime minister shortly after he was sworn in, showed it to him, and told him that we intended to conduct an operation to search the facility. He recognized the volatility of the tape, and asked me to hold off until he had some time to confer with his advisors and the presidency council. I agreed. After a few days, he told the Ambassador and me that he had decided that the public release of the tape at this time would have such a negative impact on the security situation that he was not going to release it.

He also asked me not to conduct any operations against the facility. I reluctantly agreed (largely because we had seen carloads of material being removed from the facility in the intervening days). It was a troubling indicator so early in his tenure.

The other pressing issue was the visit of the Iranian foreign minister to Baghdad. By that time, we had strong evidence of the support that Iran was providing to the Shia militia. I thought it was important for the prime minister to have access to that evidence before the meeting. The Ambassador and I had Major General Rick Zahner, my deputy chief of staff for intelligence, lay out the evidence, which he did in such a compelling way that the prime minister commented at the end of the briefing that what the Iranians were doing was conducting terrorism in Iraq. We asked him to press the Iranian foreign minister to halt the support.

We had the first meeting with the newly inaugurated prime minister and the outgoing security team a few days after the inauguration. Because I wanted to have something ready to go for the new government, I had MNF-I review the lessons of Operation *Scales of Justice* and prepare a plan to secure Baghdad. As we presented the plan, it became immediately clear that the prime minister was not comfortable with accepting a plan that he and his advisors were not familiar with, especially without his own security team on board. It was critically important that the new government make the plan their own so that they felt responsible for its execution. We agreed to have a combined Iraqi-coalition team study and wargame the plan and report back to the prime minister. It would be a few more weeks before Prime Minister al-Maliki got his security team on board and approved the plan.

It was a difficult time for the new prime minister, and the Ambassador and I worked to help him deal with the challenges he was confronted with while continuing to press for the actions that we knew were needed. It was a delicate balancing act. Civil-military interaction is difficult in any country, even in peacetime. Cross-cultural civil-military interaction in the middle of a war is even harder and requires patience and trust on all sides.

About 2 weeks into the prime minister's tenure, he was able to gain consensus on his security team—the same day, as it turned out, that our special operations task force tracked down and killed Abu Musab al-Zarqawi, the longtime leader of al Qaeda in Iraq. The task force had been tracking him for some time and finally got a break. It located him in a house a few kilometers northwest of Baquba, about 30 kilometers north of Baghdad, and attacked the house with joint direct attack munitions, destroying the house and killing al-Zarqawi and nine members of his inner circle. We were able to get a Special Forces team to the house just as Iraqi police, who had responded to the explosion, were loading a body that turned out to be al-Zarqawi into an ambulance. The team took the body and brought it to Balad Airbase for identification. DNA samples and fingerprints were taken and sent to the United States for positive identification.

We made a decision to keep the operation very quiet until we were certain of the identification. We also wanted to ensure the new prime minister had the opportunity to announce the death. Shortly after the arrival of the remains in Balad, I got a call from LTG Stan McChrystal, who had seen the remains and confirmed that it was in fact al-Zarqawi. Based on that information, I called the Secretary of

Defense and the Ambassador (I had already told General Abizaid), and we began planning for a morning announcement, pending positive identification. While we knew that this would be only a short-term setback for al Qaeda in Iraq, it was a significant success for our forces and for the new Iraqi prime minister. The next morning, Prime Minister al-Maliki announced al-Zarqawi's death to the Iraqi people at an emotional press conference. Coupled with the announcement of the new security ministers, it was a rare good day in Iraq.

Camp David and June D.C. Consultations

As we worked to get the new government on board, we were also preparing for a major U.S. conference on Iraq policy at Camp David that would be followed by a surprise visit to Baghdad by President Bush to reinforce and energize the new government. The Ambassador and I were asked to give our strategic assessment of the situation and to lay out our campaign plan for the way ahead via secure video teleconference. We described the security situation as "complex as it's been" because of the diversity of the violent groups that were now impacting security: Sunni extremists, Shia extremists (some directly supported by Iran), and the Sunni resistance. This made the term *the Insurgency* (which to date had been used to refer to the Sunni Arab rejectionist insurgency) less useful. We stated that, in the aftermath of the Samarra bombing, the fundamental nature of the conflict had changed from an insurgency against the coalition to a struggle among Iraq's ethnic and sectarian groups for political and economic power in Iraq. To us this meant that enduring strategic success would only be achieved by Iraqis as they sought to enhance unity, improve security, and build prosperity. We laid out our campaign plan to grow the capabilities of the Iraqi government so that by the end of the 4-year tenure of the new government, it would be able to govern without our assistance. Because of the shift in the nature of the conflict and the added complexity that came with it, and the ground that was lost in the 6 months of government formation, we estimated that the new government would be hard-pressed to demonstrate substantial progress in the next 6 months, and that it would take us until early to mid 2007 to stabilize the security situation to the point that political and economic development could take place without significant disruption. We also estimated that it would take us until the end of 2009 to achieve our campaign endstate.[5]

We highlighted the continuing growth of the ISF in numbers and capabilities (while acknowledging the lagging development and sectarian issues with the police forces), the success we had been having against al Qaeda leadership and facilitation networks, and the inroads we were beginning to make with the Sunni resistance leaders. There was no doubt that we faced a tough 6 months, but there was some potential for advancement *if* the government of national unity could begin to bring the country together. To help move them in this direction, the Ambassador and I suggested the development of political "benchmarks" to replace the UNSCR timeline as a forcing function to drive the political progress essential to move the country forward. The benchmarks would include such important milestones as the amendment of the constitution, provincial elections, modification of de-Ba'athification policies, and militia disarmament. We had seen over time that, as difficult as the political issues were, if there was nothing to drive Iraqi leaders to make decisions on key issues, decisions would not get made and the country would not come together politically. If, as we had postulated, the struggle in Iraq was over the division of political and economic power, we needed a political alternative to violence as a means to resolve political issues. Unfortunately, we were not able to overcome the objections of the sovereign government of Iraq. It would be a significant void in our efforts throughout 2006 and beyond.

At the conclusion of the Camp David meeting, President Bush clandestinely departed the United States and flew to Iraq. He had private meetings with the Iraqi leaders, most importantly Prime Minister al-Maliki, and then met with the new cabinet as a group. His message was one of empowerment and support, and it had the intended energizing effect on the new prime minister and his government. While there, the President also met with and thanked the members of the Embassy and MNF-I, which was a great morale boost for the team.

About 2 weeks later, I returned to Washington to serve as the president of an Army promotion board for two-star generals, to get face-to-face feedback from the Camp David session on our planned way ahead, and to take some leave as I entered my third year in Iraq.

In my Pentagon discussions, I reemphasized the points that I had made at Camp David—namely that the sectarian violence had significantly complicated the security environment, but violence, though high, was not widespread (14 of the 18 provinces had less than nine reported incidents of violence/day); that it would take about 6 months to see if the new government could make a difference; and that army development remained on track, but that we needed a major effort to restore confidence in the police. I projected that we would be finished with the planned training and equipping of the ISF by the end of 2007, that we would have all of the Iraqi divisions leading operations by the next spring, and that we would have the Iraqi provinces responsible for their own security by the end of 2007. I believed that with the government of national unity in place, steadily improving ISF, and a still-substantial coalition presence, we could move Iraq forward in the next few years *if*—and it was a big if—the new government could begin to reconcile the divergent interests of Iraq's ethnic and sectarian groups.

My discussions with the Secretary of Defense and Joint Chiefs also focused on whether or not the time was right to continue the drawdown of U.S. forces. We had done a quarterly review in March and concluded that there had been insufficient time to assess the impact of the first off-ramp decision and that there was just too much uncertainty in the outcome of the government formation process to continue to draw our forces down at that time. Having recently completed our June assessment,[6] and discussions with corps and division commanders, we had concluded that we could reduce our forces by another three brigades (about 10,000 people) over the course of the year and still maintain an appropriate level of security because of the continued development of the ISF. (In June, for the first time, we had almost as many Iraqi brigades and battalions leading operations as we had coalition brigades and battalions, 87 and 91, respectively.) We would retain the reserve in Kuwait and a force that could deploy from its home station on short notice to hedge against the uncertainty of the coming months. As Prime Minister al-Maliki and I had not completed our discussions on the proposal, a decision was put off until after my return to Iraq.

The question that we all wrestled with was why draw down coalition forces in the face of increasing sectarian violence. My thinking was that since the fundamental problem in Iraq was over the division of political and economic power, and that this conflict was the root cause of the sectarian violence, the ultimate solution would be political and not military. Furthermore, my experience had been that the longer we remained there in force, the more the Iraqis relied on us to solve their problems, and the less they moved forward on their own. I found this to be true at both the political and military levels. Finding the right force levels that would provide the right capacity for security, reduce Iraqi dependency on us, and foster resolution of the political issues at the heart of the violence was definitely more art than science.

I calculated that the specter of continued coalition reductions would reinforce the notion that the coalition was eventually leaving and create a sense of urgency in the new Iraqi government and its security forces that could spur the reconciliation that was so desperately needed for Iraq to go forward. We would still have a coalition force of over 120,000 and two of the three brigades on a short string to deploy to Iraq if we had miscalculated. I felt that I was taking a calculated risk that would take advantage of the seating of the constitutionally elected unity government to produce the substantial payoff of early reconciliation.

Unfortunately, as I was having these discussions in Washington, al Qaeda in Iraq launched retaliatory suicide attacks following al-Zarqawi's death, setting off a chain of events that made continuing the drawdown of our forces impractical in 2006, and I canceled the projected drawdown shortly after I returned to Iraq to begin my third year in command. With spiking sectarian violence and an unproven new government, I knew we were in for a tough 6 months.

Prime Minister Nuri al-Maliki and General Casey sign documents returning operational control of Iraqi forces to government of Iraq, September 2006
AP Photo (Khalid Mohammed)

5. THE TOUGHEST DAYS (JULY 2006–FEBRUARY 2007)

As I returned to Iraq at the end of June, I weighed our challenges and opportunities. After two governments in 2 years and a protracted government formation period for the third, we finally had an Iraqi constitution and a permanent and democratically elected Iraqi government based on that constitution. It was, at least nominally, a government of national unity, and it would need some time to establish itself, particularly with the sectarian tensions that it faced in the aftermath of the Samarra attacks. We continued to make progress with the ISF. The Iraqi army had held together fairly well through the sectarian violence, although we were starting to see the impact of political influence—political leaders threatening military leaders who took action against members of their factions—on the willingness of the army to take operational risks. The national police's effectiveness had been limited by absenteeism, lack of leadership, and, in some cases, direct involvement in sectarian violence. They would have to be completely reorganized to be effective. The local police were a mixed bag and still lagged the army in development. The initial operation that we mounted in Baghdad with the new government, Operation *Together Forward*, had, from its mid-June start, reduced violence in general and in the five areas of Baghdad where the sectarian violence was the worst. Unfortunately, al Qaeda had lashed out in late June with a series of suicide attacks that would continue into July, driving retaliatory attacks by Shia death squads that further inflamed the situation. The combination of the two—suicide attacks and death squad executions—had led to a spike in violence against civilians that we would have to contain. The situation was further complicated by continuing evidence of Iranian training and equipment support to Shia militias.

Adjusting the Plan

The first session that I had with my staff on my return was my monthly intelligence update. It was a sobering brief on the security situation, particularly around the capital. My intelligence officer highlighted the recent attacks by al Qaeda and the Shia death squad backlash and assessed that the tit-for-tat violence had become self-sustaining. He stated that he was beginning to see an almost predictable cycle of al Qaeda suicide attacks, followed in a few days by Shia death squad attacks against Sunni areas in Baghdad. The sectarian violence was focused in Baghdad and southwest Diyala Province and, in those areas, focused more on civilians than on Iraqi and coalition security

THE TOUGHEST DAYS

forces.* The violence across the rest of the country remained relatively low, with the exception of Anbar Province where the violence was primarily directed against the coalition and not sectarian. In fact, around 80 percent of the violence in Iraq continued to remain centered in 4 of its 18 provinces: Baghdad, Anbar, Diyala, and Ninewah. This finding was reinforced in the unit visits that I undertook in the first week following my return. The other disturbing finding was that my analysts were beginning to see a geographical component to the violence—that the Shia death squads may have been trying to drive Sunni families out of mixed neighborhoods to improve their control of Baghdad. If this were true, it would mark another worrisome shift in the conflict. It would bear careful watching. As I left this briefing, I began to rethink the plans to reduce our forces that I had discussed in Washington.

The other major focus in early July was to refine the Baghdad security effort that we had begun in mid-June. Because of the need to get the new government to act quickly in the face of rising sectarian violence and the late appointment of the security ministers, the new Iraqi leadership had not participated in planning the initial operation to the degree that they desired. As they gained more experience, they wanted more of a role. I saw this as a positive step in that they were willing to begin to take ownership over the plan to secure their capital.

So, while continuing with Operation *Together Forward*, we began working with the Iraqis to enhance our collective efforts in Baghdad to bring security to the capital by the end of the year. I told the MNC-I commander that Baghdad was our main effort and that he needed to

---

* From May to mid-July 2006, 40 percent of the violent incidents in the Baghdad area were against civilians, 32 percent against the ISF, and 28 percent against the coalition.

develop a plan to secure Baghdad that was sufficiently weighted to ensure our long-term success. He began working with Iraqi military and police planners and crafted a plan based on an operational concept where joint (Iraqi and coalition forces) would clear areas of Baghdad of enemy control and then protect these areas while we improved the capacity of the ISF, and worked with the Iraqis to improve services (electricity, water, sewage) in the areas. MNC-I would simultaneously limit al Qaeda and death squad movement by creating a barrier around Baghdad linked to the canal system and channeling all traffic into Baghdad through checkpoints. In addition, the Iraqis would manage a system of fixed and mobile checkpoints around the city to further limit extremist mobility. MNC-I would also conduct targeted offensive operations against death squad and al Qaeda targets in Baghdad and in the surrounding support zones. This plan was worked painstakingly with Iraqi leaders to ensure that we had their buy in and strong commitment to its success. We had originally proposed beginning with clearing operations in Sadr City, but the prime minister did not support this action, and we began with a focus west of the Army Canal that divides Baghdad. The next phase of Operation *Together Forward* was approved for implementation by the prime minister and me in early August, and it began shortly thereafter.

As we worked through these adjustments, several conflicts that would hamper our efforts became clearer. I began to see that the prime minister and I had fundamentally different views of the threat. I felt that the Shia militias were the greatest threat to our ability to bring security to Baghdad and to long-term security in Iraq. I showed Prime Minister al-Maliki data that the casualties from death squads (largely Shia) far eclipsed the casualties from the more spectacular suicide attacks (largely al Qaeda). The prime minister believed that the "Ba'athists" (Sunni extremists) were the greater threat, and he expressed concern that we were putting all of our efforts against the militia and not enough against the Sunni extremists. I tried to counter this notion by having our special operations task force commander show him the scope of our significant effort against al Qaeda. He also believed that the Shia militia could be dealt with politically, but that the "Ba'athists" could only be dealt with by force. This was the reason he turned down our request for a major operation into Sadr City, where we thought the main militia threat was coming from. The prime minister also did not see the "geographical component" of the Shia violence, further impacting his reluctance to deal expeditiously with the militias. We would work on reconciling these views over time, but, as they were strongly held by both of us, they would cause increased friction between us as we wrestled with bringing security to Baghdad and Iraq. I note these conflicts to demonstrate the complexities of conducting military operations inside a sovereign country and the importance of political and military leaders having a common view of the threat to drive effective military operations.

Canceling the "Off-ramp." By mid-July, we had concluded that our long-held assumption that the government of national unity would be seen as representative by most Iraqis and have a positive impact on the security environment was not going to hold true. The new government was seen as not representing the interests of a good portion of the population. The differences in threat view that I saw were also visible to Sunni leaders and only compounded their negative views of the government. It had also become clear that the Iraqis would not be able to secure their capital without more support from us.

In the weeks since my return from Washington, DC, given what had transpired in Iraq between mid-June and mid-July and the negative way the government was being perceived by the Sunni population, I realized that there would be no strategic payoff from drawing down coalition forces at this point, and that we needed to focus everything we had on securing Baghdad. This meant that we would have to forego the planned off-ramp of coalition forces that I had discussed in Washington in June.

As I was reversing myself from my June position, I wrote a personal, classified message to General Abizaid (whom I had kept abreast of my changing thinking), Chairman of the Joint Chiefs of Staff General Pace, and Secretary Rumsfeld. I described how the situation had changed since mid-June—the significant spike in violence against civilians that we saw in late June and early July, the reluctance of government leaders to limit the actions of militias and death squads, and the increasingly geographical nature of Shia death squad actions. I also related my discussions with the prime minister on the reductions—he was reluctant to see us take reductions now with the levels of violence so high.

I told them that I needed to keep more coalition troops in Iraq than I had expected to help the new Iraqi government contain the increasingly difficult situation, and that I needed a "full-court press" on the political side to jumpstart the reconciliation process as a complement to our security efforts. I asked to keep a brigade that I had intended to send home without replacement in order to reconstitute the reserve in Kuwait (we were bringing the current brigade forward to augment the Baghdad plan) and to establish a force ready to deploy on 30-days notice at home station in case of larger problems. I stated I

THE TOUGHEST DAYS

would need these additional forces through the end of the year. I concluded by restating my belief that, while the extra forces would help the security situation in the short term, they would not have a decisive impact until Iraqi religious and political leaders committed to stop the sectarian killing.[1] I was concerned that without a commitment by the Iraqis to a reconciliation process, our continuing resolution of the security problems would allow them to postpone the reconciliation that was essential to our collective long-term success.

The request was approved expeditiously. This effectively canceled any further plans to reduce our forces through the end of the year,* although we did continue with our plans to pass security responsibility to Iraqi provinces as their security capabilities and local security situations warranted. The first Iraqi province to assume responsibility for its own security, the southern province of Muthanna, did so on July 13, 2006.

After we canceled our plans to reduce our forces, and as we continued our planning to secure Baghdad, the MNC-I commander approached me with an option to request a 90- to 120-day extension for a redeploying Stryker brigade to give us a mobile strike capability in Baghdad. After initial reluctance because of the turbulence caused by extensions, I agreed that the potential operational benefit in Baghdad at this critical time would be significant and requested the extension.

In retrospect, I waited too long to make the decision to cancel the drawdown and to extend the Stryker brigade, and this caused substantial turbulence at the tactical level. We had a deliberate process in place that we went through in June with our commanders in which the recommendation to off-ramp three brigades in 2006 was made.

---

* In September, I revised this through the spring of 2007.

What we did not have in place was a deliberate process to revisit the decision as the situation changed visibly. In the end, it was a combination of unit visits, interactions with Iraqi leaders, and conversations with my subordinate commanders and staff that led me to change my mind. Reversing yourself is hard to do, especially when you have publicly committed yourself to a course of action, but it is something that every leader in war will have to do. The sooner you do it, the better.

Looking Ahead. Over the course of my command, I tried to create an environment where we asked ourselves hard questions and challenged our assumptions. It was the only way to stay ahead of the complex and constantly changing situation.

So before I left for Washington in June, I had formed two Red Teams, one to examine ways to counter Iranian influence in Iraq and the other to see if we needed to rethink our strategic priorities in the face of rising sectarian violence. We had done reviews of the situation and our priorities in April with our commanders that led to our presentation at Camp David. These two Red Teams were designed to move our thinking beyond that level.

The team on Iranian influence concluded that the best way to counter Iranian influence was to counter the operations of the Islamic Revolutionary Guard Corps–Quds Force, the action arm of Iranian interference in Iraq. We had good intelligence that they were providing training and modern equipment to Shia death squads, most notably the explosively formed penetrator, a particularly lethal IED that we had begun to see in large numbers in the Baghdad area in late 2005. What we did not have was targetable intelligence to go after their operatives. To rectify that, we established an Iran fusion cell, a multidisciplinary intelligence collection and analysis center exclusively dedicated to countering malign Iranian influence in Iraq. As with all new intelligence operations, it took some time before it was producing actionable targets. At that point, I aligned the cell with our special operations task force to action the targets. The effort paid off handsomely in December when we caught several Quds Force operatives and confirmed a lot of what we suspected about Iranian activity.

The second Red Team reviewed our 2006 action plan in light of the recent violence and noted that while our priority in Baghdad was the correct strategic priority, it was "inadequately resourced across all lines of operation" and that the new government did not have the capacity to secure Baghdad without significant coalition support over the next 6 months, conclusions which played in my decision to cancel our planned troop reductions.

I reviewed the second Red Team assessment with the staff in late July. They generally agreed with the findings of the team, and I directed them to look across all of the lines of operation and determine what additional resources should be moved to Baghdad. Our original plan called for us to prioritize Baghdad and nine key cities.* We would keep moving forward outside Baghdad, but we would accept delays in other places in order to focus key resources on Baghdad.

In mid-August, the Ambassador and I convened our key staff and commanders to review the situation and to ensure that the new staff from the summer rotation understood the direction that we planned to head over the next 8 to 10 months. We laid out our priorities. Our first priority was to mass all of our efforts—military, political, economic, and informational—to secure Baghdad. Second,

---

* We had refined our list of 15 key cities to 9 after the first election.

we had to sustain country-wide pressure on al Qaeda and the death squads to keep them out of Baghdad. Third, we needed to sustain progress "away from the ball"*—continuing to develop the ISF, especially the police; continuing to transfer security to Iraqi provinces that were ready to assume security responsibility; continuing our work with the PRTs to build capacity at the provincial level; and continuing with economic development around the country. In short, we needed to continue to execute our campaign plan where security conditions permitted, while we worked with the Iraqis to secure Baghdad.

The plan to enhance security in Baghdad was christened Operation *Together Forward II,* and it began in earnest in early August with the planned addition of 12,000 Iraqi (unfortunately, the Iraqi troops failed to arrive) and coalition troops to the Baghdad mission. The additional troops included five military police companies that would work as transition teams with the Iraqi police to shore up their staying power and evenhandedness—the two major issues we had with the local police. The plan also began a retraining program for the national police where a brigade at a time was pulled offline, refitted with equipment and retrained, and its leadership purged of sectarian influence. The program produced positive results over time.

With Operation *Together Forward II*, we sought to make a demonstrable improvement in the Baghdad security situation by the beginning of Ramadan (September 23–October 22). We expected to

---

* I used a basketball analogy—"play away from the ball"—to make the point that securing Baghdad was a long-term proposition, and, as important as that was, it was not the whole country. We had to keep making progress in the rest of the country—"away from the ball"—while we worked to improve security in and around Baghdad.

see increased violence during Ramadan as we had every year.* This year the expected increase in violence was helped by a September 7 call by the leader of al Qaeda in Iraq, Abu Ayyub al-Masri, to kill an American in the next 15 days.

Clearing operations in the focus areas, the areas of highest sectarian violence, proceeded well, and violence decreased from July levels through the end of August. We intended to maintain momentum by expanding efforts to additional focus areas, completing the Baghdad barrier to limit extremist freedom of movement, consolidating our gains in the focus areas by improving security and basic services there, and maintaining pressure on al Qaeda and death squad leadership. As we cleared the focus areas, the terrorists and death squads shifted their efforts outside of the cleared areas and continued their attacks. By Ramadan, the operation had kept attacks against civilians and civilian casualties below July levels, but had not stopped the sectarian killings.

We continued to make gradual reductions in the sectarian violence and to keep the pressure on the al Qaeda and death squad leaders through Ramadan, but coalition forces bore the brunt of the violence. Most of the casualties were the result of IEDs, and almost 85 percent were in Baghdad or Anbar provinces. September was a difficult month, particularly because we were beginning to see indications that the Iraqi security forces were not performing to standard. Moreover, in some cases, particularly in the national and local police, we found active collaboration with the militia. Additionally, we were seeing very slow responses by the Iraqi government to bring services into the Sunni areas that had been cleared and in moving Iraqi brigades into

---

* Historically, we had seen violent incidents increase by 15 to 20 percent during Ramadan.

Baghdad in support of the plan. There was also little movement on the Iraqi political front in support of the security efforts.

Civil-Military Relations

The lack of Iraqi political support to security efforts continued to be disappointing. In the types of operations we were conducting in Iraq, political and military actions are irrevocably linked. I strongly believed that until the Iraqis began to resolve the political issues that were dividing them, the rationale for violence would not be eliminated, and Iraq would continue to struggle, requiring the continued presence of coalition forces. I believed that they were capable of resolving these issues because I had seen them work together in the tough days before the January 2005 elections.

Throughout the late summer and early fall, there was a concerted U.S. effort to get the Iraqi government to agree to a series of "benchmarks" designed to establish a timeline for the resolution of the difficult political issues dividing the country. Issues such as reviewing the de-Ba'athification process, establishing a timeline for provincial elections, reviewing the constitution as promised before the December elections, and completing the hydrocarbon law needed to be resolved in order to better align political and economic power in the country—the issue at the heart of the violence. We believed that equitable movement on these issues would cause the government to be seen as more representative of the entire Iraqi population and ultimately lead to a lessening of the violence. Unfortunately, we lacked sources of leverage to move the Iraqis to action. They were a sovereign country with a duly elected government, and their leaders intended to exercise that sovereignty. They tended to act on their timelines and assessments, and the leaders that I dealt with, while conscious of their security responsibilities, did not see the situation as being as serious as it was believed to be in the United States and coalition countries. These differing views created friction and tensions in our relationships that required continuous attention. They also caused some on my staff to wonder whether the current government of Iraq and the U.S. Government had a common vision for the future of Iraq—a troubling question.

I continuously worked to maintain a professional working relationship with all Iraqi political leaders. Building the relationship with Prime Minister al-Maliki was not easy for either of us, and it had its ups and downs because of our different cultural backgrounds, the inherent difficulty of civil-military interactions during war, and the fact that we were working on different timelines and responding to different constituencies. Going back to my earliest days in Iraq, I firmly believed that the Iraqi government had a better chance of being viewed as legitimate by Iraqis than the coalition, so we worked hard to make every prime minister successful. I felt that this was especially important with the new government, as they would be the ones who would ultimately guide their security forces to secure their country.

Prime Minister al-Maliki and I got off to a rough start when, in retrospect, I pushed too hard to get started on improvements to the Baghdad security plan in the early days of the new government. The prime minister came in with strong views on the ISF and the utility of force (he thought 3 years of coalition use of force had hurt more than it helped) that would take some time to reconcile, and events seemed to inject friction into our relationship almost daily. For example, at a meeting in August, he objected to reports in the U.S. media suggesting that he ordered attacks on Sunni but not Shia targets. He was concerned because it was not true and that it had reached President Bush. He wanted to know if it had come from me. It had not. I told him that I suspected that it had come from people below me who were upset with constraints on our operations in Sadr City.

At another meeting in September, where the deputy chief of the U.S. mission and I laid out the options for the next phase of operations to secure Baghdad, the prime minister told me that he thought I was second-guessing his decisions on military operations. I told him that I did not think that I was. He emphasized strongly that he made decisions based on his convictions about what was best for the country and not along sectarian lines. These were difficult sessions, but I was glad that he thought enough about the importance of the relationship to speak openly about what was on his mind. I came to realize that until we resolved our differing views of the threat—his seeing Ba'athists as the most dangerous threat, and my seeing the militia as the most dangerous—we would continue to be at odds. I continued to work to address those differences.

Toward the end of October, one of our soldiers was kidnapped in downtown Baghdad. The division commander reacted quickly and established checkpoints throughout the city to recover the soldier. We kept the checkpoints in place while we ran down every lead. We had kept these checkpoints in place for a week when we heard that the Sadrists were pressuring the prime minister to remove the checkpoints. I called the division commander and asked if the checkpoints were still necessary. He stated that they were no longer necessary for finding the lost soldier, but that they had seen a general drop in violence during the week that the checkpoints were in place. The prime minister called the next day and asked for a meeting.

The Ambassador and I went to his office, and he was clearly agitated. He felt that we were inconveniencing a city of 8 million people to find one soldier. I told him that one soldier was important to us. He stated he wanted us to remove the checkpoints. I told him that if we did that, he could be seen as not caring for coalition forces, caving in to the Sadrists, and caring more about Shia than Sunni Iraqis, who were benefiting from the increased security of the checkpoints. If he was willing to accept those consequences, I told him that I would remove the checkpoints during the day, but keep them in place at night. He said that he accepted the consequences, and I instructed the division commander to open the checkpoints during the day. All of the consequences I predicted happened, but I felt that it was important for the new prime minister to understand that there were second- and third-order effects of his decisions.

I highlight these tensions for future senior leaders. Interaction between military and civilian leaders is always difficult in war. It is even more difficult with leaders from other cultures and countries. Trust is the important commodity in these relationships, and frank and open dialogue is the only way to maintain it. While the frictions in our relationship would continue, the prime minister and I worked hard to maintain an open dialogue.

During this time, I also had significant civil-military interaction with my own government. In early October, I returned to Washington to discuss plans for the next year. My key concern was how we were going to stop the sectarian violence and move the country forward. I told different administration audiences that the violence was hardening sectarian divisions within the country and that we needed a political track along the lines of the benchmarks that were being worked to complement the security effort. There were "chicken-egg" discussions about whether the security situation had to improve before the political track could begin. I strongly argued that both tracks needed to move forward simultaneously to be effective, and that it was important to get the Iraqis to commit to a political timeline—the benchmarks.

I also laid out our projections for 2007 that were conditioned on getting movement on the political side. We projected having all 10 of the current Iraqi divisions in the lead and under the operational control of the Iraqi Ground Forces Command by the spring. We projected having all Iraqi provinces responsible for their security by the fall. We also projected completing the planned development of the ISF by the end of 2007, less the national police retraining and the completion of the prime minister's initiative to expand the armed forces and police. Prime Minister al-Maliki's initiative, which he directed shortly after taking office, added 50,000 soldiers to the army (an additional 2 division headquarters, 3 brigade headquarters, 12 battalions, and a 10 percent manning overage for existing units to offset absentee issues). It also included the reform of the national police mentioned previously and development of a national counterterrorism force. This would be a substantial enhancement to their capabilities. It would take until the late summer of 2007 to complete police reform and until the end of the year to complete the prime minister's initiative. As I had informed the Service chiefs that I would need the forces I had through at least the spring of 2007, there were no discussions of additional troop reductions. I cautioned that we were in a difficult period and that the risks remained substantial.

On my return to Iraq, as we continued to address the security issues in Baghdad and contain the Ramadan surge in violence, I visited the coalition divisions and some of their brigades to get a first-hand view of how things were going on the ground. With the exception of the units in the Baghdad area, most were reporting continued progress. I was particularly pleased with a visit to Ramadi, the capital of Anbar Province, where the new unit had made significant progress since August, pushing al Qaeda out of the city and implementing the security force plan for the province that the previous prime minister had approved. The combination of our military actions against al Qaeda and the hiring of security forces from Anbar, vetted by their tribal leaders, was galvanizing the tribes in Anbar against al Qaeda.

At the end of October, President Bush and Prime Minister al-Maliki agreed to establish a working group to accelerate the pace and training of the ISF, the Iraqi assumption of command and control over the ISF, and the transfer of security responsibility to the Iraqi government. As we worked with the Iraqi security leadership and our subordinate units to operationalize this agreement in November, we were adamant that it had to be accompanied by an Iraqi-led reconciliation effort to be successful. I thought that might be our leverage—if the Iraqis wanted to advance the time when they would control their security forces, they would have to take the steps toward reconciliation that were essential for our collective, long-term success.

The Ambassador and I suggested an integrated political-military framework to demonstrate how political actions and security actions could be mutually reinforcing and lead to our long-term success (see figure 5-1). The framework began with the Council of Representatives (COR) debating and passing a series of laws that would codify the division of political and economic power—hopefully in a manner that represented the interests of all ethnic and sectarian groups. This would be followed by reconciliation and militia disbandment efforts that would culminate in an amnesty agreement that would be implemented over a period of time. This agreement would be followed by provincial elections and a referendum on amendments to the constitution to make it a truly national compact. The entire political process would be supported by security and stabilization operations that would establish security in Baghdad and nine other key cities and the ultimate assumption of security responsibility by the ISF. What we tried to replicate was the type of political timeline that we had through the end of 2005 that drove Iraqi actions. We were unsuccessful in gaining acceptance of it by the Iraqis.

On November 8, the President accepted Secretary Rumsfeld's resignation from his position as Secretary of Defense in the aftermath of the midterm election where Republicans lost control of both the Senate and House. While he agreed to stay on until his replacement was confirmed, this reinforced to all of us the deep dissatisfaction with how the war in Iraq was going. The administration's Iraq Strategy Review that I had heard about in October kicked into high gear, leading to the most complex period of my command. In the last 2 months of 2006, we were simultaneously revising and intensifying our efforts in Baghdad, planning for 2007, working with the prime minister and his security team to eliminate malign political influences on the security forces and to build Iraqi political support for the Baghdad mission, dealing with Washington's concerns about the direction of the mission, and managing the transition of a new corps.

Military Operations in Baghdad

In November, we began adapting the Baghdad security plan to deal with the evolving threat. With the exception of the Ramadan

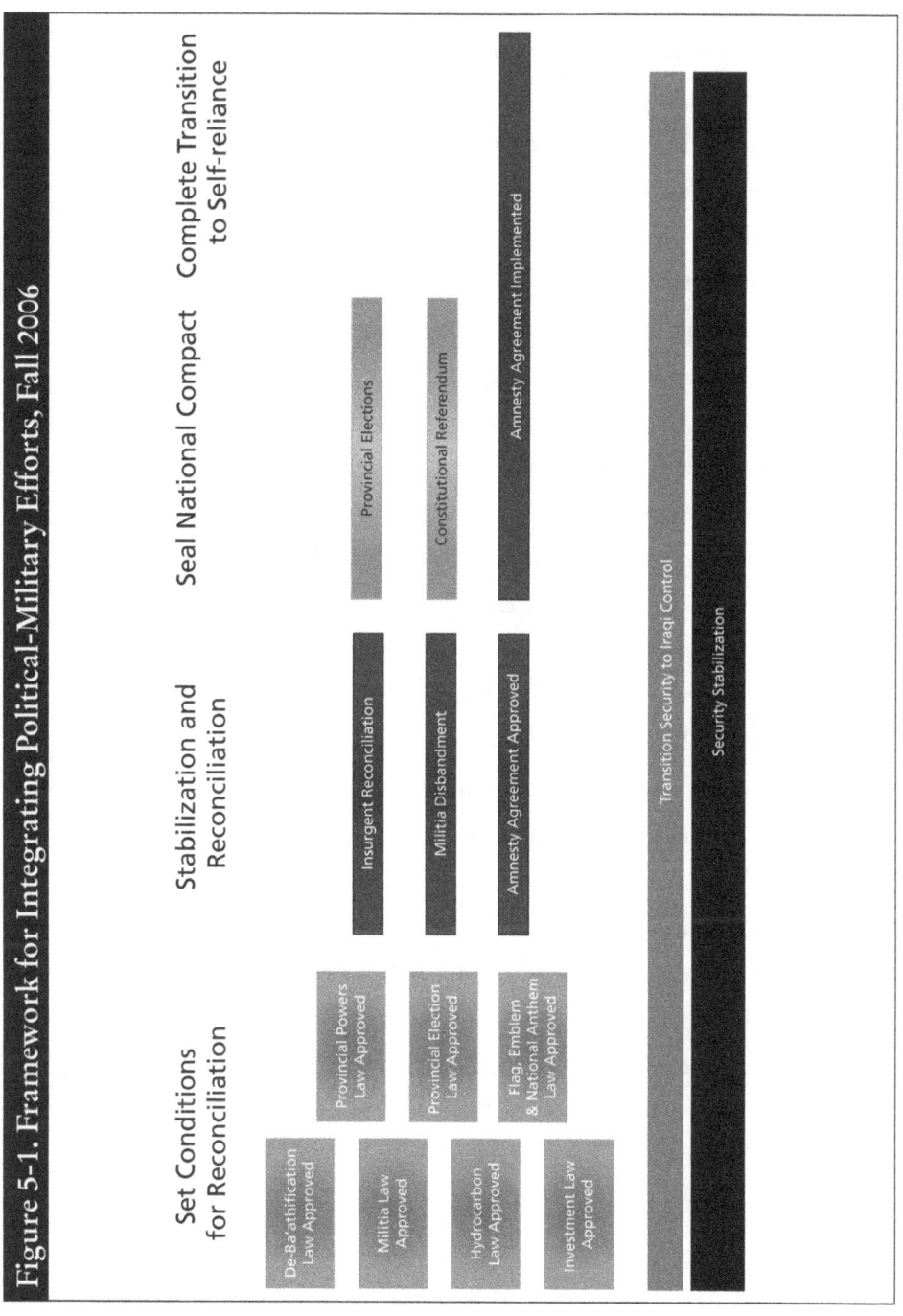

Figure 5-1. Framework for Integrating Political-Military Efforts, Fall 2006

increase in violence, we had kept violence against civilians and casualties below July levels, and we were holding steady in the cleared areas, but the level of violence was still unacceptably high.

In one of our early November security meetings, the prime minister and security ministers had just returned from a closed session with the Council of Representatives on the security situation. It was a seminal meeting in that all of the Iraqi security team spoke their views forcefully and openly. Interestingly, the prime minister reported that the prevailing COR view was that the problem in Iraq was not security, but political. He stated that the COR agreed that the ISF were free to operate anywhere in Iraq and would issue a statement to that effect—good news if it held up. He claimed that there was increasing evidence that the Ba'athists were the greatest danger and expressed concern that they could influence the Iraqi army. I countered that there were two sides killing civilians in this fight and what was required was a balanced security effort against all the armed groups. The prime minister agreed that all threats should be attacked, but said that some could be better dealt with politically and others with force. Our different view of the threat was still there.

At the end of the meeting, the prime minister laid out a list of things he wanted his ministers to do, and clearly empowered them to conduct operations against all militant groups—the militia, Ba'athists, and terrorists—something that was sorely needed. He also authorized us to conduct operations into Sadr City against the indirect fire teams that were targeting the Green Zone and sections of Baghdad. It was good to see this sense of urgency.

The Iraqi leadership continued to frame their thinking over the intervening days, and when I met the prime minister 2 weeks later, he

THE TOUGHEST DAYS

was clearer in his intent. He wanted to make a national statement that no one was above the law and that anyone acting against the government would be subject to swift action from the Iraqi security forces. He planned to obtain political support from all of the major parties for this in advance. He then wanted to follow this up with major Iraqi-led operations against both Sunni terrorists and Shia militia to demonstrate the strength of the ISF and the commitment of the government to end the violence. I was impressed with his energy and commitment.

Unfortunately, the next day six car bombs hit Sadr City killing or injuring more than 400 people, sending us into the crisis action mode. The security team responded quickly as they sought to limit retaliation and prevent another incident. As the crisis passed, work on the broader security plan for Baghdad continued.

As we were also in the throes of another MNC-I transition, I wanted to use the fresh eyes of the new leaders and the experience of the departing leaders to improve our efforts in Baghdad. Securing a city of 8 million people that was about the size of "inside-the-Beltway" Washington, DC, was no easy task, and we continually looked for ways to improve our security efforts. The Baghdad division changed out on November 15. I immediately sat down with the new division commander and his corps's commander. We told the division commander to take a blank sheet of paper and craft a plan, in conjunction with the Iraqis, to secure Baghdad. We gave him 30 days to report back with a plan and what he would need to execute it.

I had also directed the outgoing MNC-I commander to put together a 120-day assessment of corps's efforts in Baghdad. He gave us the assessment before he departed on December 14. While he pointed to the successful tactical efforts to clear areas of Baghdad and to target death squad and al Qaeda leadership that had mitigated but not substantially reduced the levels of sectarian violence in the city, he was clearly frustrated at our inability to sustain our tactical successes. He reported that, in general, the clearing operations were successful, but the Iraqis lacked the staying power and sometimes the will to hold the cleared areas after coalition forces left. The Iraqis were also very inconsistent in delivering services to the cleared areas. He pointed specifically to:

- inability of Iraqis to deliver promised units to Baghdad
- Iraqi political constraints on military action, most notably in Sadr City where the bulk of the death squad activity came from
- militia infiltration of security forces, primarily police
- lack of policies limiting militia and the carrying of weapons
- ineffectiveness of the Baghdad barrier system because of unprofessional and corrupt conduct at the checkpoints by the ISF
- political interference with ISF leaders.

I had gotten the same sense of frustration from the Baghdad division commander and the Diyala brigade commander before they left. They did not see the Iraqi political support necessary for our security efforts to take hold. They felt as if we wanted Baghdad secure more than the Iraqi leaders. This was troubling as it threatened our partnership with the Iraqis, which was at the core of our efforts to secure Iraq. I resolved to address this with the prime minister before we launched the next iteration of our Baghdad efforts.

The report was not all negative. Curfews, driving bans, and Army Canal checkpoints proved effective. The Iraqi army was proving adept at holding areas, and the Iraqis were getting better at coordinating army, national police, and local police operations through a joint command center that they had established. All of these were necessary to move forward.

We blended our lessons with the proposals made by Prime Minister al-Maliki and the Iraqi security ministers. There was general agreement that there were too many stationary checkpoints tying up Iraqi forces and that these needed to be reduced to free up forces for offensive operations. There was general agreement that the ISF needed to be on a more offensive footing like the coalition, patrolling and conducting targeted operations against terrorists and death squads. The Iraqis also perceived a need for more joint operations—coalition, army, police—as a means of building trust between the Iraqi army and police forces and suppressing the likelihood that any Iraqi forces would succumb to sectarian influences. They suggested joint security stations across Baghdad, located in selected local police stations, where coalition, Iraqi army, and police forces would be based and operate out of to bring security to the surrounding areas. Finally, we worked with the Iraqi leaders to establish an Iraqi command structure for Baghdad. While I had originally envisioned a joint coalition/Iraqi command for Baghdad, the Iraqis convinced me that they were capable of taking command, still with our support. They were taking ownership of the mission to secure their capital, which was essential for our success.

The operational structure established for Baghdad was composed of two commands, one east and one west of the Tigris River that almost bisected Baghdad, under a Baghdad operational command, headed by an Iraqi three-star who would report directly to the prime minister. Each of the commands would be joint army/police commands, one (east of the river) commanded by an army division commander with a police deputy and the other (west of the river) commanded by a police two-star with an army deputy. Each of the nine Iraqi districts in Baghdad would come under the command of an Iraqi army or police brigade augmented by the local police assigned to the district and a coalition battalion. This structure made sense from a military perspective, and it allowed the prime minister to better fix responsibility on his Iraqi military and police leaders. It also lessened the likelihood of the Iraqi forces becoming involved in sectarian violence as everyone—army, police, and coalition forces—would be watching each other.

The plan called for five additional brigades to be moved to Baghdad (three Iraqi and two coalition) and the execution of a phased effort to establish long-term security. In the first phase, 35 joint security stations would be established and occupied by Iraqi and coalition forces. This was a significant logistical and construction effort that we estimated would take around 6 weeks to complete. In-place Iraqi and coalition forces would continue their security efforts to sustain pressure on the extremists during this phase. Then areas would be cleared, expanded, and held by the joint forces, and, over time as violence lessened, the ISF would assume full responsibility for the security of their capital. The fact that forces would flow into Baghdad over time actually helped us in that the new forces ensured that we could continue to hold and expand cleared areas—something that had eluded us in the previous efforts.

THE TOUGHEST DAYS

On December 23, the minister of defense briefed the plan to Prime Minister al-Maliki. He laid out the operational concept mentioned above, and recommended to the prime minister that the military operations be accompanied by robust media, political, and economic reconstruction campaigns to ensure sustained political support for the operations. He recommended closing Iraq's borders for a period of time to limit the entry of external threats into the country during this critical period. He also recommended that the prime minister address a commanders' conference to provide his intent directly to the army and police commanders who would be executing the plan. On the question of timing, always difficult in civil-military discussions, the MOD estimated that it would take about a month from the prime minister's approval before the nine-district command structure would be up and functioning (which I thought was optimistic by a few weeks). He said that operations in two of the districts would begin right after the first of the year and that sustaining operations would also continue across Baghdad while the command structure and bases were established. We had jointly pressed to kick off offensive operations with the New Year. Unfortunately, the movement of forces and construction of the joint security stations would take longer than we had hoped. We agreed that we should not announce the start of a big operation early on, but rather to begin the operations, construct the joint security stations, flow in additional forces, and let the people see the accomplishments of the forces.

The prime minister approved the plan right before Christmas, and I clarified with him that he was also approving the deployment of two additional coalition brigades in support of the plan. He acknowledged that he was, but was clear that he wanted to downplay the significance of the deployment of additional coalition forces. This left two remaining issues—the appointment of the commander and the prime minister's speech in which he would publicly empower the security forces to take action against "all who broke the law." Both of these would be resolved after the first of the year.

For the commander, Prime Minister al-Maliki chose a relative unknown, at least to the coalition, Lieutenant General Qanbar Abud, who had worked directly for the prime minister, and was clearly someone whom he trusted. I was initially uncomfortable having someone so unknown to us responsible for our main effort, and I had several sessions with the prime minister where I expressed my concerns. In the end, he appointed him, telling me that if Abud proved that he was not up to the job, I should let him know and he would replace him. In the end, the general proved a capable choice.

In his Army Day speech on January 6, the prime minister laid out the key points we had been looking for from him to empower his security forces. He stated that the government would:

- "Not permit any political authority to weaken our armed forces . . . because weakening the army will lead to delaying the process of receiving the security responsibility from the MNF"
- "Not allow any militias, regardless of their belongings, to be a replacement for the state"
- "Enforce the law against all those who infringed the sacred rights of the Iraqi people."

With respect to the Baghdad security plan, he reemphasized that the government would not tolerate political interference, that the security forces would pursue all "outlaws, regardless of the sectarian or political affiliation," and that military commanders would be given "all authorities to execute the plan." He closed by calling on the people of Baghdad to support and assist their armed forces. We, and our Iraqi military and police colleagues, finally felt that we had the political backing that had been lacking to pursue both the Sunni and Shia extremists who were fomenting sectarian violence. It was a good start.

Washington Policy Review

As we were wrestling with the tactical and operational challenges brought on by sectarian violence and the change in the nature of the mission, Washington was grappling with its strategic implications. I was informed about a review of the Iraq strategy in October by General Pace, but, from my perspective, it did not begin in earnest until after Secretary Rumsfeld's resignation in early November.

As part of this process, I met with the Iraq Study Group, a congressionally appointed, bipartisan commission, by video teleconference in early November. I had spoken to the group in Iraq in August, and they asked to speak to the Ambassador and me again before they finalized their report. I made five points with them:

- ◆ Conflict in Iraq is about the division of political and economic power among Iraqis.
- ◆ Sectarian violence is the greatest threat to the accomplishment of our strategic objectives. Reconciliation among Iraqis is essential for our success.

STRATEGIC REFLECTIONS

- Enduring strategic success will only be achieved by Iraqis—and it will take longer than we want even to get to "Iraqi OK."*
- We are two-thirds of the way through a three-step process to bring the Iraqis to the point where they can credibly assume security responsibility by the end of 2007.
- We have adapted and adjusted our strategy, plans, and troop levels to meet the changing dynamics of the situation on the ground.

Their questions were good ones: Why wasn't the Baghdad security plan having greater effect? Is the prime minister the right guy, and is he willing to go after the militias? Should we transition without reconciliation? I answered these questions and told them that we were in no danger of losing militarily, and that more coalition troops would have a temporary and local effect on the security situation. I also commented that more coalition forces at this point would give Iraqi leaders more time to avoid hard decisions on reconciliation and ultimately prolong our time there.

The report was released on December 6 and offered 79 recommendations advocating internal and external approaches to reverse what they called a "grave and deteriorating" situation. I was heartened to see the co-chairs note in their opening letter that there were no "magic formulas" to solve Iraq's problems and that no one could "guarantee" that any course of action would work to stop the sectarian violence. Externally, they recommended a diplomatic offensive

---

* I used the term *Iraqi OK* to make the point that we were trying to get Iraq to a level that was acceptable to Iraqis, and not imposing U.S. or European standards on them.

to build international consensus for stability in Iraq and the region. Internally, they recommended that the United States "adjust its role to encourage the Iraqi people to take control of their own destiny." On the military side, they recommended accelerating the assumption of security responsibility by Iraqis and changing the primary mission of U.S. forces to one of "supporting the Iraqi army." They also recommended that the United States work closely with Iraqi leaders "to support the achievement of specific objectives—or milestones—on national reconciliation, security and governance"—something that we had been trying to establish for months with the benchmarks.[2]

I also had visits from the National Security Advisor and his deputy at the end of October and early November, respectively. As I did with most visitors, I did not accompany them as I felt that through our frequent video teleconferences, they knew what I thought. I wanted them to hear from my subordinate leaders without any impression of influence. I did review their itineraries to ensure where they went enabled them to meet their trip objectives and I met with them to answer their questions.

In mid-November, shortly after my video teleconference with the Iraq Study Group, I was informed that the Deputy National Security Advisor was working to get the strategy review to the President before the end of November. A week later I sat through a video teleconference with the Joint Chiefs to discuss their independent strategy review to help General Pace shape his military advice for a meeting with the President later that week. Their proposal was to accelerate passing the security lead to the Iraqis. It was based on two big ifs—achieving unity of effort with the Iraqi government and the government making progress on reconciliation. It proposed shifting our main effort to training and partnering with the ISF, which was the approach that we were working.

While it did not seem to me to be connected to the proposed strategy, we also discussed sending five additional brigades to Iraq by mid-April for 3 to 9 months to help get the security situation under control. I told the Joint Chiefs that we could certainly put the additional forces to use and that they would have a temporary, local effect where we put them, but questioned the impact that the additional forces would have on the Iraqis' incentives to resolve their differences. I felt that the longer we remained responsible for their security (the impact of the additional forces), the less incentive they would have to resolve their own differences, which was essential to our long-term success. I also cautioned that, if we did send them in, we should not accept any limits on their employment from the Iraqi government, and require progress on reconciliation as prerequisites for bringing in additional forces. We needed to get something substantial from the Iraqis for such a significant additional expenditure of U.S. forces. In his feedback from the session with the President, General Pace told me that there had been no decisions, but to expect a decision on the new way ahead by the third week in December.

On November 30, President Bush and Prime Minister al-Maliki met for the third time, this time in Amman, Jordan. In a video teleconference a few days before, Mr. Hadley discussed the objectives for the meeting with the prime minister, noting that it was an opportunity for al-Maliki to explain his intended direction for the unity government to the President so that he could support him in his efforts. Mr. Hadley proposed three topics for the delegation meeting: an update from the prime minister on the situation in Iraq, a U.S. update on the

Baker-Hamilton Iraq Study Group, and a report on the joint committee for accelerating the transfer of security responsibilities that the prime minister and President Bush had agreed to the month before. The prime minister agreed to the proposed agenda and during the intervening days also had his staff draft a proposal for securing Baghdad that contained many of the proposals that we had been discussing with the ministers. The important element was that it was his plan. Although it was not what we would think of as a military plan, it was a good overview of the policies and principles that he saw as important to succeed in Baghdad, which we could incorporate into the ongoing, collective coalition and Iraqi military planning efforts.

I took advantage of a few minutes with the President before the meeting to update him on the three topics and encouraged him to reiterate that the prime minister must have militia reintegration and reconciliation strategies for us to proceed with the accelerated transition programs, and to support the prime minister's proposed security plan for Baghdad. During the meeting, the two leaders exchanged views on the situation, and al-Maliki discussed his Baghdad plan and asked for the President's support to help it succeed. The President agreed. They also discussed the work of the joint committee and agreed that any accelerated transition was dependent on the security situation, reconciliation, and militia reintegration. The meeting concluded with a joint press conference in which the President reiterated what he and the prime minister had talked about. President Bush went on to call the meeting with the prime minister "a key part of the assessment process"—the review of Iraq strategy.

The first week in December, the pace of the review picked up substantially. It began with a video teleconference with the NSC on

STRATEGIC REFLECTIONS

December 1, in which I gave an update on the current situation. To demonstrate that the country was not "aflame," I began with a slide that showed that only four of Iraq's provinces (Anbar, Baghdad, Saladin, and Diyala) had averaged more than 10 reported attacks per day in the 6-month period from May through November, the height of the sectarian violence. I showed another slide showing that only two of Iraq's provinces (Baghdad and Diyala) had averaged more than three sectarian casualties per day during that same period. The rest averaged less than one per day.

Our problem was Baghdad, and it was significant. Most striking was the fact that we were averaging more than 40 civilian casualties per day in Baghdad over that period and between 150 and 250 deaths per week as a result of sectarian violence. This was clearly unacceptable. I also showed that the vast majority of civilian fatalities were caused by the largely Shia death squads as opposed to the suicide attacks of the Sunni extremists. I closed with a slide that showed how we had substantially increased our operations against both Sunni extremists and the death squads since the summer to demonstrate the level of offensive action we were taking. In November alone, we had killed or captured over 800 extremists and death squad members in targeted operations.

In early December, General Pace called to say that my session to provide input to the President on the strategy review would be on December 12. It would be preceded by a session with outside experts on the 11th and followed on the 13th by a meeting with the President and Joint Chiefs. He expected a decision and the announcement of the results before Christmas. This was not my only session during the review. In the week of December 8–15, we had five video teleconferences with the NSC. The issues covered ranged from how to ensure the support of the Iraqis, to the potential size and composition of a "civilian surge" to accompany a surge in military forces, to how to deal with a confrontation with the Sadr militia, to how to enhance operations in Anbar Province. It was the policy process at work—a meeting with the NSC, questions, scrambling for answers, staff meetings to prepare for the next meeting of the NSC, another NSC meeting, and a repetition of the cycle. It was all focused on framing the issues and providing the best information possible to permit the President to make the most informed decision. In the end, there was general agreement that success in Iraq was essential to our national security and that, while reconciliation was essential, there would have to be a reduction in sectarian violence to allow reconciliation to take place, which would, in turn, provide for a more stable longer term outcome.

My session on December 12 included General Abizaid. I used the opportunity to lay out my proposed way ahead. I proposed an integrated political-military effort to stabilize the country and pass security responsibility to the Iraqis in 2007 as had been agreed in Amman. I was clear that the Iraqis would still require coalition support beyond 2007 and that the level of that presence should be negotiated with the Iraqis over the course of 2007. I stated that accomplishing what I proposed would require coalition forces to:

- assist the ISF in quelling sectarian violence and neutralizing the extremists
- support Iraqi efforts on reconciliation and dismantling militias
- complete the training and equipping of the ISF by the end of 2007

- continue our efforts against al Qaeda, death squads, and Iranian surrogates.

We would also continue to work with the Embassy to build Iraqi institutional capacity at the national and provincial levels and to continue economic development.

I also highlighted the risks involved, which were not insignificant, and more on the political than the military side. I worried primarily about the ability of the Iraqi leadership to take the necessary political steps to support our security operations—reconciling the interests of the different ethnic and sectarian groups, dealing with militia and illegal armed groups, giving our forces free rein to attack hostile targets, and eliminating political interference with the ISF.

I knew there was a push to move five U.S. brigades into Iraq to deal with the security situation. I had asked for two to meet the needs of the Baghdad security plan and two battalions of Marines to maintain our momentum in Anbar Province, so I offered my thoughts on bringing more forces than that into Iraq. I stated that additional forces:

- would have a temporary, local effect in reducing sectarian violence where they were committed
- could provide breathing space for a committed government to address militia and reconciliation challenges
- would place the new forces in a complex environment where consent for their presence was diminishing
- could extend the time it takes to pass security responsibility

- would result in additional coalition casualties
- would not have a decisive effect without government commitment to reconcile and deal with the militias.

After watching the impact on our Baghdad division of having to operate in such a complex, politically constrained environment, I was very concerned about bringing fresh U.S. troops into the middle of a sectarian conflict in an Arab country where there was not clear political support for their actions. I felt very strongly that I would not ask for one more American Servicemember than needed to accomplish our mission, especially in this environment.

This was an intense period as it was clear that Washington was looking for something different from what I was recommending to them. I worked hard to provide my military advice dispassionately, as I felt that I was providing the President only one of the many options he was reviewing. I believe that a President is best served by having a variety of options to choose from. I had said all along that success in Iraq would take patience and will and believed that what I was recommending—to accelerate the transition of security to capable Iraqis in exchange for their action to solve the core problem in Iraq, that is, reconciling the interests of the different ethnic and sectarian groups—offered the most effective way to accomplish our strategic objectives in Iraq. I believed that I had asked for the troops that I needed to accomplish our operational objectives, and that, if the prime minister delivered on his pledges to the President to allow our forces and the ISF to operate freely without political interference, we would bring security to Baghdad by the summer. I felt that additional troops beyond that would risk introducing them into a very confusing and difficult operational environment without a plan for how their introduction would contribute to the accomplishment of our strategic objectives. I remained adamantly opposed to that.*

In retrospect, I believe that I should have directly offered the President a broader range of options for achieving our objectives in Iraq. I had discussed different options for improving the security situation with the Secretary of Defense and Chairman: accelerated transition of security responsibility; local (with in-country forces), small, and large coalition reinforcement; coalition withdrawal on a fixed timeline; and maintaining the status quo. Only the accelerated transition and reinforcement were actively considered. In the end, I only presented the President the course of action we selected—accelerated transition—and I believe that I should have offered him a wider range of options to meet his policy needs.

The pace kept up in the weeks before Christmas. MNC-I swapped out on December 14, bringing a new team into the complex environment. Secretary Rumsfeld departed on the next day after coming to Iraq earlier that month for a farewell visit. Secretary Robert Gates took over 3 days later and made his first visit to Iraq as the Secretary of Defense on December 20 with General Pace. General Abizaid and I met with them and laid out our views on the situation. Secretary Gates was familiar with the issues as he had been sitting in on the video teleconferences during the strategy review. We also continued to work with Iraqi leaders to finalize the plan to

---

* While we did our own course of action analysis in December and studied the logistical implications of bringing in the additional forces, we had ideas, but no operational plans, for the additional three brigades. These would be developed by MNC-I in February and March before the forces flowed into Iraq.

THE TOUGHEST DAYS

secure Baghdad until we received the prime minister's final approval on December 23.

In the middle of all this, our efforts to target Iranian operatives paid off with the capture of six Iranians who appeared to be engaged with Iraqi militia in planning for the expansion of Shia-controlled areas in Baghdad. It was the first time that we had clear evidence of this. Four of the six had ties to the Iranian embassy and were released in a few days. We believed that the other two were Quds Force operatives who had entered Iraq under false names and had no right to diplomatic status, so they were held as we continued to evaluate the material that was discovered with them—maps, weapons receipts, and money. The most disturbing element was that they appeared to be working very closely with Badr Corps operatives. The Badr Corps was a militia with close ties to one of the main Shia political parties.

I continued my Christmas tradition of visiting the troops and thanking them for their work before heading back to Washington for some face-to-face discussions. During my session with General Pace at the end of December, he informed me of major pending decisions on the Iraq strategy by the President and his national security team. Specifically, he told me that our "2 + 2" proposal (two brigades for Baghdad and two battalions for Anbar) had been judged as "too modest," and that, while there was not yet a final decision, he expected one by the end of the month that would add a total of five brigades and supporting forces. We took some time so he could be clear on the difference between my request and the likely Presidential decision.

I was asking for the two brigades that we needed to implement the Baghdad plan and two Marine battalions to maintain our momentum in Anbar Province, about 9,000 troops. We expected the first brigade to flow in by mid-January, the second by mid-February. The additional three brigades, if approved, would flow at the rate of one per month, if they were required. (I knew that I would be leaving shortly and wanted to give my successor as much flexibility as possible by having the option to turn off deployments if he decided he did not require the additional brigades.*) A few days later, the Chairman informed me that the President had decided on the five-brigade surge and that the President intended to nominate LTG Dave Petraeus to replace me. I had provided the President my military advice on what I felt was the best approach to accomplish our strategic objectives in Iraq as rapidly as possible. He chose a different course of action. His decision was disappointing to me, to say the least, but I immediately set out to make it successful.

As Washington prepared for the rollout of the "surge" strategy, we were working hard to set the conditions for its success and to finalize the plans for securing Baghdad. This included a video teleconference between President Bush and Prime Minister al-Maliki on January 4 to ensure that they shared a common understanding of the new Baghdad security plan and that the prime minister was prepared to provide the political support for the coalition and Iraqi forces that was essential for the success of the plan.

During the video teleconference, the President informed the prime minister of his inclination to increase coalition troop levels provided that they reach "a common understanding." The President

---

* Army Chief of Staff General Pete Schoomaker told me that I was being considered to replace him in the spring, and Secretary Gates confirmed this during his visit to Baghdad. Secretary Gates offered me the Army chief's job, which I accepted before I returned to Iraq in early January.

was frank, stating that the additional coalition forces were meant to help the Iraqis break the back of terrorism to help accelerate the transfer of responsibility to the Iraqi government. He noted that the United States was willing to commit to help secure Baghdad, but that Iraqi commitment was also very important. He told Prime Minister al-Maliki that he needed him to publicly state his government's commitment prior to the President's planned address to the American public on January 10. The prime minister was cautious and judicious in his responses, noting that it was important they work together. He stated that his cabinet would start planning and would get back to him in several days. On January 6, Iraqi Army Day, Prime Minister al-Maliki gave his promised speech outlining the elements of his Baghdad security plan in which he strongly made the points that President Bush had requested.

On January 10, the President outlined his decision in a prime-time speech that announced a plan "to help the Iraqis carry out their campaign to put down sectarian violence and bring security to the people of Baghdad." He announced the commitment of 20,000 more troops to Iraq and that the majority of them—five brigades—would be deployed to Baghdad. The remainder would go to Anbar Province to "work with Iraqi and tribal forces to keep up the pressure on the terrorists." He couched the mission in Iraq in broader terms, calling our struggle against extremism in the Middle East "the decisive ideological struggle of our time," and stating that the new plan would "change America's course in Iraq, and help us succeed in the fight against terror."[3] It was a moving speech and a powerful statement of U.S. commitment to Iraq.

In a press conference that the Ambassador and I held in Iraq a few days later, I noted that the plan to secure Baghdad had several key advantages, most important of which was the strong commitment of the Iraqi government, including the will to act against all who broke the law and not to impose restraints on the ISF and coalition forces. I discussed how the plan allowed us to sustain the agreement in Amman between Prime Minister al-Maliki and President Bush to accelerate development of the ISF and the passage of security responsibility. The introduction of the additional forces allowed us to sustain momentum, reinforce success, and evaluate progress as we went. I emphasized the flexibility inherent in the plan and how it was a strong statement to the Iraqi people of our commitment to securing Baghdad and accomplishing our mission in Iraq.

My final task in Iraq was to ensure the conditions were set for the new Baghdad plan to succeed. We worked to establish joint security stations that would house Iraqi army, police, and coalition forces that would bring security to Baghdad's neighborhoods. We worked with the Embassy to establish the funding and mechanisms for follow-on economic projects in both Baghdad and Anbar. We worked with the Iraqis to finalize command and control arrangements, finalize the selection of the Iraqi commander for the Baghdad operation, and establish the Baghdad operations center to give the new commander a headquarters. We also developed the logistical support plans to receive, equip, support, and base the incoming coalition forces. I met frequently with the Iraqi ministers to review their preparations and personally reviewed the final plan to ensure that it adequately incorporated the projected influx of forces.

In a January 20 video teleconference with President Bush and Secretary Gates, I updated them on the ongoing preparations. The first brigade to flow from the United States was closing in Baghdad and was beginning operations. Offensive shaping actions that we had begun in Baghdad after the first of the year—some 14 battalion or larger operations in 20 days—continued to put strong pressure on al Qaeda and the death squads through daily intelligence-based raids. The Iraqi command and control headquarters—Baghdad and the two sectors—were expected to be operational in about 2 weeks. About one-third of the joint security stations were operating, with almost 20 more projected to come on line by mid-February. Iraqi army brigades were being alerted and moved on the agreed timelines, but were arriving at between 55 and 65 percent strength, and the Iraqis were moving to address the shortfalls. In all I was pleased with the progress that we were making and the sense of energy that I was seeing in the Iraqis. I was also pleased that we had seen a 5-week decline in sectarian violence and civilian casualties in Baghdad and that we continued to make good progress in Anbar.

In the middle of all this, I returned to Washington on February 1 for my confirmation hearing to be the Army chief of staff. I flew straight into a Pentagon "murder board" and 2 days of office calls with the members of the committee. The hearing was a tough one as the Senators asked hard questions about my 32 months in Iraq. I was confirmed on February 8, 2007.

General Petraeus had been confirmed on January 27, and we set the date for our change of command for February 10. When I returned to Iraq after the hearing, I focused on our continuing preparations in Baghdad and on setting the conditions for a smooth turnover with General Petraeus.

At my final meeting with Prime Minister al-Maliki, I offered my thoughts to him on civil-military interaction with his military and police uniformed leadership. I had told him once, early on, that he and

I were like two old men in a rowboat. If I pulled on the military oar and he did not pull on the political oar, we went around in circles (and vice versa)—but if we both pulled together, we moved forward together. I felt very strongly that effective political-military integration was the key to long-term success in Iraq, so I offered him eight tips for civilian leaders in providing guidance to military leaders that I had translated into Arabic:

- Be clear with them what you want them to accomplish. This will become the mission.
- Make them tell you how they see the enemy situation. You should have a common view of what you think you are up against.
- Ask them for their assessment of planning and preparations. Key questions are: How long until you are ready? Do you have what you need to do the job? How long will it take?
- Make a collective judgment of the appropriateness of the plan relative to the mission and threat.
- Ask about the plan for reaction forces if things do not go according to plan.
- Ask about the risks of mission accomplishment, collateral damage, friendly casualties, and adversely affecting broader objectives.
- Ask them to explain the command and control relationships. Be sure they are clear.
- Ask what help they need on the political and economic side.

THE TOUGHEST DAYS

Before we parted, the prime minister gave me a copy of the Iraqi constitution signed by him and his security ministers, and I gave him the pistol that I had carried throughout my 32 months in Iraq.

Two days later, on February 10, 2007, I relinquished command to General Petraeus. In my remarks, I commented on how far Iraq had come since it achieved its sovereignty over 2½ years ago and expressed my deepest gratitude to the Servicemembers and their families who had given up so much to build a new Iraq and bring liberty and democracy to 27 million Iraqis. I closed with the Arabic words, *Iltizam Mushtarak* (United Commitment), which had been the motto of Iraqi and coalition forces during my time in Iraq. I was very conscious of the difficult challenges still facing the mission, but I felt that I had done everything possible to set the conditions for our ultimate success.

It had been a long 32 months, but I believe that the efforts of the men and women who served in Iraq during that period drove a significant transformation in the U.S. military and established the conditions for the ultimate success of our mission in Iraq. The completion of the UN political timeline that led to an Iraqi constitution and the seating of an Iraqi government based on that constitution in just 24 months established Iraq as a democratic state after over 3½ decades of totalitarian rule—a significant historical accomplishment. The growth of the Iraqi security forces from a relative handful of army battalions and police forces to a force of over 325,000 that was actively participating in securing their country and had held together during difficult sectarian violence is a tribute to the men and women from over 30 countries who trained, mentored, and fought beside them. This growth established the necessary Iraqi security capabilities to begin the process of transferring the ISF and provinces to Iraqi control, our ultimate mandate from the United Nations. By the end of January 2007, the Iraqi ground force headquarters, 3 of 10 Iraqi divisions, and three provinces (Muthanna, Dhi Qar, and Najaf) had been returned to Iraqi control. The One Team/One Mission concept and the integration of the Embassy and MNF-I staffs enabled us to build a structure to integrate, synchronize, and assess the progress of the U.S. mission. Its success is a tribute to the professionalism, competence, and dedication of our Foreign Service officers, intelligence service professionals, and Armed Forces who worked hard to break down institutional biases to get the mission accomplished.

Men and women of our Armed Forces and our allies liberated Iraqis from decades of oppression. They succeeded and found themselves enmeshed in a conflict and an environment for which they were not prepared. They improvised, learned, adapted, shared their lessons, and, over time, improved our capability to operate effectively. From countering the IED threat to fundamentally reshaping detention and interrogation operations to revamping contingency contracting procedures to training the Iraqi army and police, they led a transformation of the way in which the U.S. military prepares for and conducts 21st-century conflict. Our success is a tribute to their courage, their perseverance in the face of adversity, and their professionalism.

General Casey at Senate confirmation hearing for Army chief of staff, February 2007
AP Photo (Susan Walsh)

6. INSIGHTS FOR LEADERS

I have thought a great deal about my experiences in Iraq. I believe that some of the insights that I developed during that time can benefit future military leaders as they are thrust into senior leadership positions in new and different missions in this era of persistent conflict. As always, some lessons are new; others are old ones relearned. I began to share these insights with the Army general officer corps and joint flag officers attending CAPSTONE shortly after I assumed the position of Army chief of staff.

Perhaps the greatest lesson I took from my time in Iraq was that senior leaders are most effective when they stay at the right level and focus their time and intellectual energy in the areas that will yield the highest payoff for their organizations. That sounds easy, but it is not because the things with the highest payoff are

153

the hardest to do—for example, getting the strategy right in very uncertain environments; instilling the strategy in the organization; driving organizational change; influencing organizational culture; sustaining momentum; and influencing key partners not under your direct control. By their nature, these things are complex and difficult and do not lend themselves to simple solutions. They require the time, energy, and experience of the senior leaders in the organization to be done effectively. What follows are some insights in those areas for future leaders.

Developing Vision and Strategy

The question that I asked most in Iraq, and, interestingly, the one I asked most as Army chief of staff, is, "What are we really trying to accomplish?" I found that this question was hard to answer clearly and succinctly in the complex and uncertain environment of Iraq. Yet it was imperative that I clearly articulated to my subordinates what it was I wanted them to do if we were going to be successful. A fuzzy idea coming out of the four-star headquarters did not get clearer as it was transmitted through the chain of command. Accordingly, we spent a lot of time and intellectual effort sharpening our views of what we wanted to accomplish in Iraq and for major operations, for example, Fallujah, elections, the western Euphrates campaign, and Baghdad.

The Army's primary doctrinal manual, Field Manual 3-0, *Operations*, offers a construct to assist commanders in framing solutions to difficult problems—understand, visualize, describe, direct—and, although we did not think of what we did in those terms at the time, that is what Ambassador Negroponte and I did initially as we grappled with the mission. We both felt that we needed to establish a clear vision for what we were to accomplish in Iraq, so we began discussing it before we left Washington. The consultations that we conducted in Washington, the Red Team assessment, and the on-the-ground consultations in Iraq in the early days after our arrival were all part of building our *understanding* of the mission. As our understanding grew, we began to sharpen our thinking on what we wanted to accomplish and how we wanted to accomplish it—we began *visualizing* the endstate and design for the mission. In interactive discussions with Washington, the Red Team, the Iraqis, and our staffs, we began *describing* how we saw the mission unfolding and received their insights. In dealing with the complexity and uncertainty of Iraq, I found that building a level of understanding sufficient to visualize the problem and to describe the solution effectively was an iterative process—that my thinking got sharper over time. I found that the sharper the disagreements, the greater the clarity we achieved.

We gave ourselves 30 days to produce a joint mission statement and campaign plan, the means by which we would *direct* the tasks required to accomplish the mission. We felt strongly that we owed our subordinates as much clarity as possible to shape a common path to success. I found it particularly important to be clear on the nature of the war we were fighting—counterinsurgency—and the nature of the enemy—primarily Sunni Arab rejectionists—and to clearly spell out the mission and the risks. I felt that the 30-day timeframe was important because I had seen too many draft campaign plans that were continuously being polished and never published. In complex situations, commanders must force themselves to get clarity in their own minds and transmit that clarity to their subordinates in writing.

I found that writing things out caused me to think more clearly about issues, so I personally wrote several of the key segments of the first campaign plan (for example, mission, intent, risks).

We built a deliberate assessment process into the campaign plan because we knew the plan would require continuous adjustment. As part of this process, we forced ourselves to challenge our assumptions and ask ourselves hard questions about the efficacy of the plan. The assessments proved useful in adapting our efforts to changing realities. I also found there was constant tension between retaining focus on the broader campaign and adapting to short-term changes in the environment. One of the ways that we used to mitigate this tension was to publish annual campaign action plans that allowed us to retain the focus on our broad counterinsurgency campaign while dealing with shorter term issues. The annual action plans also proved helpful in maintaining continuity through the transition of subordinate units and staffs.

I am convinced that one of the hardest things for leaders to do in complex and uncertain environments is to get clarity in their minds on what it is they want their subordinates to accomplish to achieve success. Because it is so hard, it takes the full involvement and commitment of the senior leader to accomplish it successfully.

Creating Unity of Effort

Another difficult challenge for senior leaders is to create unity of effort among organizations whose cooperation is necessary for their success, but that are not under their direct control. The National Security Presidential Directive issued in May of 2004 established the division of labor between the Departments of State and Defense for the mission

INSIGHTS FOR LEADERS

and directed "the closest cooperation and mutual support" between the Ambassador and the USCENTCOM commander. United Nations Security Council Resolution 1546 described my relationship with the soon-to-be sovereign government of Iraq as a "security partnership." If we were to successfully prosecute a counterinsurgency campaign inside a sovereign country, I was going to have to rely heavily on the Embassy and Iraqi government to deliver the political, economic, communications, and, in the case of the Iraqis, security effects to support coalition efforts. The keys to my success were outside of my direct control, so I was forced to create the required unity of effort with successive Ambassadors, Iraqi prime ministers, and cabinet ministers.

Ambassador Negroponte and I recognized this early on and agreed before we left Washington on the One Team/One Mission concept—the Embassy and MNF-I would work as one team to accomplish the U.S. mission. Because of different organizational cultures and different reporting and budget chains, implementing the concept took the direct intervention of the Ambassador and me. Conscious of the need to bring the missions together intellectually, we established a Red Team composed of key leaders from both organizations to tell us what they thought about the mission and the threat. Putting a key advisor to the Ambassador as the leader of the effort and giving him a strong military deputy allowed us to get a balanced output from the group. The Red Team report led to the joint mission statement by the Ambassador and me that was a key step in establishing One Mission. The essence of the statement was dutifully incorporated into the campaign plan so that it penetrated MNF-I. The Ambassadors and I issued joint mission statements three times during my time in Iraq as the mission evolved.

Building the "One Team" was equally challenging. The old adage that "Defense is from Mars and State is from Venus" just scratches the surface of the cultural differences between two professional communities. Given human nature, major institutional and cultural differences do not disappear in a war zone, and working through them requires the continuous involvement of senior leaders. The Ambassadors and I went to great lengths to bring the two organizations together and keep them moving in the same direction to accomplish our national goals in Iraq. We used the Red Team concept frequently to keep us intellectually aligned. We collocated our offices, traveled together, and consulted regularly and visibly to ensure our subordinates saw us linked together. We integrated our headquarters with the Embassy to provide the physical proximity necessary for effective coordination. Sustaining the One Team/One Mission concept between the Embassy and MNF-I took a lot of the personal time and effort of the Ambassadors and me, particularly with the annual rotation of staffs and two changes of Ambassador.

Over our initial weeks on the ground, the Ambassador and I wrestled with the implications of Iraqi sovereignty on our efforts. The United States had returned sovereignty to the Interim Iraqi Government on June 28 and the Coalition Provisional Authority had appointed Ayad Allawi as the interim prime minister. We recognized that unless we shared our vision and plans with the Iraqi leadership, we would not only generate unproductive friction between us, but also be unable to leverage the influence of the government in support of our efforts. While the Iraqi government had publicly accepted MNF-I presence, the modalities of coordinating our operations had to be worked out. We set out to establish them

INSIGHTS FOR LEADERS

in a way that respected Iraqi sovereignty but that retained our freedom of action. Sovereignty meant that the Iraqis had a vote and that things would not necessarily get done the way we wanted when we wanted. The Ambassador and I would have to balance Washington's directives and timelines with the needs and desires of the sovereign Iraqi government. It was a delicate balancing act, and one that required our almost constant attention. I cannot overstate the benefit we got from spending the time to establish strong personal relations with Iraqi leaders. Strong personal relationships can help bridge the frictions that will always be encountered.

My staff and I found that we spent a lot of time integrating the efforts of the Embassy, three Iraqi governments, and MNF-I. There were frustrating days when I asked myself whether this was the best use of our time. In the end, I saw it as my headquarters' responsibility to work with the Embassy and the Iraqi government to deliver the political, economic, and communications effects that would make MNC-I security operations successful and sustainable. Just generating these effects in a postconflict state, let alone integrating them at the required time, was very hard work. In the end, I believe that creating unity of effort among diverse entities beyond your control is, and will continue to be, one of the key tasks that will require the attention of senior leaders in 21st-century warfare.

Continuous Assessment and Adaptation

In long missions such as Operation *Iraqi Freedom* where leaders are intensely immersed in difficult issues daily, it is easy to lose your perspective on the larger mission. I found that we had to create opportunities to get leadership to take a step back and look broadly at the mission. We built an assessment process into the campaign plan to do this, but it took some time to get it to the point where it was producing meaningful insights.

We began with a monthly assessment called the Commander's Assessment and Synchronization Board. It was designed to help us see how the staff was accomplishing the objectives assigned to them in the campaign plan so that we could make short-term adjustments. It was highly detailed. It quickly became clear to us that what we were measuring did not change that much in a month and that the staff was expending a great deal of energy developing the product, so we went to campaign assessments every 2 months.

The greatest challenge we found was determining what to measure. Staffs will tend to measure what they can, not necessarily what you need. It was not until I forced the staff to answer three questions about each of the effects we were tracking that we began getting good value out of our assessment sessions. The three questions were: What are we trying to accomplish? What will tell us if we are accomplishing it? How do we measure that?

Initially, the assessments were produced by the MNF-I staff and attended by the Embassy and MNF-I leadership. Over time, the Embassy staff got more and more involved until the assessment became a joint, and better, product. We would periodically invite representatives from the Joint Staff to attend to facilitate transparency in sharing information.

We also instituted semiannual assessments called Campaign Progress Reviews to give us a broader perspective. These reviews looked back over the past 6 to 12 months and offered recommendations for the next 12 months. They were essential to driving long-range planning. As it is a constant struggle for senior leaders to get their subordinates to share their doubts with them, I left the development of this assessment to the staff and the writing to the gifted colonels in our plans and assessment shop. I found the anonymity of the staff process produced greater candor. I found this process and product most helpful in seeing broad changes required in the mission and in developing our annual action plans. We used these assessments to adapt the mission over time. For example, the need to get better visibility on and performance from Iraqi security forces that led to the development of the transition team and partnership programs came from the December 2004 assessment. A shift in the nature of the most significant threat from former regime elements to Islamic extremists that took place in the spring and summer of 2005 and led to the western Euphrates and Tal Afar operations later that year was identified in the June 2005 assessment. The significant shift in the nature of the conflict that took place after the Samarra bombing in 2006 and that led to an increased focus on Baghdad and operations to lessen sectarian tensions later that year came from the June 2006 assessment.

There were three other forums that also enhanced our ability to adapt. The first was the monthly intelligence update where our intelligence officer reviewed intelligence trends with the staff and me. I found this forum most useful for putting the insights and thoughts that I had accumulated over the month into perspective. It allowed me to better assess the impact of individual incidents in a broader context. The second was the monthly commanders' conference where I sought to balance the MNF-I view of the mission with the views of the division and corps commanders. While I generally visited each of the divisions once or twice a month, having them share their views in a common forum proved invaluable. The third was the use of an almost continuous Red Team process to focus the attention of experienced individuals on hard topics outside of the normal staff process. I often asked the intelligence agencies to take the lead and usually included individuals from the UK embassy and intelligence services on the Red Teams. I found this process especially helpful in looking ahead (for example, I asked teams to tell me the likely outcome of elections and the implications for the mission). I found Red Teaming an excellent way to get fresh ideas and to avoid the "group think" that can often come from the staff process.

Leaders at every level must see themselves and see their enemy, and recognize that the action-reaction-counteraction cycle of war requires constant assessment and adaptation. At the theater level, I tried to focus on adjustments that would have a high payoff at my level.

Influencing Organizational Culture

At the strategic level, leaders need to be attuned not only to the culture of the country they are operating in, but also to the impacts that the cultures of their own organizations can have on their ability to accomplish their missions. I entered Iraq with views about aspects of Army and Marine Corps organizational cultures that I felt could hamper our ability to accomplish the mission if we did not address them. First, both Services are very well trained in conducting conventional war, as they demonstrated during the ground war. I knew that they would be very good at applying force against their enemies. Unfortunately, success in counterinsurgency operations requires much more than the effective application of force. I knew that it would be tough to change this mindset, but in an environment where distinguishing the enemy was very difficult and civilian casualties bred additional enemies, we would have to do it. Second, I worried that our "can-do" attitude would make it harder for us to get the Iraqis trained and responsible for their own security—the precondition of our ultimate success. I saw the impact of this attitude myself in Bosnia and Kosovo. In complex environments, it is very difficult to get even simple things done, so the natural tendency is to do them yourself. I had to find a way to get our troops to focus on Iraqi solutions without damaging the can-do spirit that sets U.S. Servicemembers apart, and that we would need to succeed.

To do this, I realized that I was attempting to change deeply embedded Service culture and that I would have to change the mindset of the force. I greatly underestimated how long this would take. We began by clearly stating in our campaign plan mission statement that we were conducting counterinsurgency operations to send the message to the force that we were doing something different than they had been trained for. I reinforced this in my discussions with leaders during their campaign plan backbriefs.

But that was hardly enough, and shortly thereafter we took measures to improve our understanding and application of counterinsurgency doctrine. We had MNF-I staff take a historical look at successful and unsuccessful counterinsurgency practices in the 20th century, and disseminated their work to the force and to Service trainers who were preparing the next rotation.

Our efforts continued with the implementation of the transition team and partnership concepts in early 2005. For the first time since Vietnam, we were asking conventional forces to be involved in the training of indigenous forces during a war—another significant cultural change. The establishment of Phoenix Academy to train all of the incoming transition team members, use of Special Forces to train conventional forces in the art of working with indigenous forces, and development of the "flat-assed rules" to communicate the new mindset to every member of the command played key roles in driving cultural change in our forces. This was a start, but we slowly began to realize that changing the organizational culture embedded in the Services for decades was not going to happen overnight.

In the summer of 2005, I chartered a survey of how we were applying counterinsurgency doctrine across the force. The study found that, while we generally knew the doctrine, it was being applied unevenly across the command, and the application was very dependent on the local commander's knowledge and initiative. It recommended that we establish a COIN Academy to augment the training that they were getting at home station to ensure that entering commanders started with a common view of how to conduct counterinsurgency operations in Iraq. We conducted the first class in November of 2005 and began to see an appreciable change in the conduct of our operations throughout 2006 as all company, battalion, and brigade commanders began to rotate through the weeklong course before they began their tours in Iraq. Continuing change was facilitated with the publishing of the joint Army–Marine Corps counterinsurgency manual in December 2006, an essential element of driving cultural change within the Services.

In the end, I found that as our lessons learned were continuously incorporated into Service training programs and more soldiers came back for second and third tours, I saw continuous improvement in the preparedness of the forces to conduct counterinsurgency operations and work with the Iraqi forces. Recognizing the impacts of organizational culture comes from the experience of growing up in the culture. Recognizing the potential impacts in new situations requires a broader perspective and is intuitive work. It is the work of senior leaders.

Civil-Military Interaction

Civil-military interaction around matters of policy and strategy is inherently challenging. The issues are complex, the stakes are high, and the backgrounds of the people involved can vary widely. The interaction only gets more difficult in war, and is particularly difficult with leaders from other cultures. Developing plans and strategies, reporting, managing expectations, and developing and providing military advice to civilian leaders all require the senior leader's full attention.

My previous experience at the policy level in Washington taught me not to expect written direction from civilian leaders, and that proved the case in Iraq. We developed the initial campaign plan based upon my verbal discussions with the President, Secretary of Defense, and Chairman, the direction provided in the President's Army War College speech, UNSCR 1546 and its attached letters, written guidance from the USCENTCOM commander, and my interactions on the ground in Iraq with Iraqi and coalition leaders. The Ambassador and I developed our strategy and campaign plan to accomplish the endstate that we created from this guidance and presented it for approval by the Secretary of Defense and President in August of 2004.

Throughout the mission, I had interaction with Washington several times a week usually in the form of secure conference calls and video teleconferences, most with the Secretary of Defense, Chairman, and USCENTCOM commander, and weekly in a National Security Council meeting chaired by the President. These sessions were designed to keep Washington up to date on the situation in Iraq. In them, I would usually present a short update and highlight upcoming events to avoid surprises. I would then answer questions. Periodically, about every 4 to 6 months, I would return to Washington for face-to-face discussions. This was essential because it is difficult to have substantive discussions on a video teleconference that includes a dozen Cabinet-level leaders with staff often operating from multiple sites. It is also much easier to get a sense of how your presentation is being received in person. The Secretary of Defense and Chairman would also visit several times a year, presenting the best opportunity for discussion and interaction. I had almost daily interaction with General Abizaid by secure telephone and face-to-face contact several times a month during his visits to Iraq or my visits to his headquarters in Qatar. His broader perspective was invaluable in seeing the Iraq mission in the context of the larger war and region.

It is difficult for subordinates to communicate to their superiors the depth of the complexity that they are dealing with. It is no different at the strategic level. I worked hard to provide a balanced view of what was occurring in Iraq—the bad with the good. I realized early on that, as I had the direct interactions with civilian leaders, I had the best understanding of what they needed, so I found that I spent a lot of my time and intellectual energy preparing properly balanced presentations for Washington. I felt that it was very important to convey the right balance in the presentations to avoid creating false expectations. I was not uniformly successful. I found it difficult to keep the discussions at a level that would provide civilian leaders with the insights they required to develop the strategies and policies essential for success. Even at the strategic level, leaders can get captivated by tactical actions.

Setting common expectations is another difficult but essential task. In any military campaign plan, it is important to set objectives and make judgments on when they will be accomplished. As senior military leaders, we owe our civilian leaders our best views on how long things will take. When we offer our views, we need to be clear that in war things will change and assumptions will prove invalid. I would often conclude a briefing in which I made key projections with a slide entitled "Bad Things That Could Happen" to make this point. When I was conveying timelines, I was very conscious that precious little in Iraq got accomplished right on time, so I would often convey projections to Washington "seasonally"—for example, we would complete a certain task by "the summer of 2006"—to give them a perspective on time without getting unnecessarily specific.

A key expectation to resolve is how to measure progress at the strategic level. Going into Iraq, we made a conscious decision not to use enemy casualties—body count—to measure strategic progress. I believe that was the right decision, but the unintended consequence was that our casualties were reported and the enemy's were not. It appeared to some domestic audiences that the enemy had the upper hand—which was not at all true. Over time, I began selectively reporting enemy losses to give a more balanced picture of the situation to our home audiences.

We looked at a variety of ways to measure progress at the strategic level, primarily focusing on significant events and milestones that, linked together, would demonstrate steady progress toward our ultimate endstate (for example, elimination of terrorist safe havens, success in major military operations, successful elections, completion of the UN timeline, seating of governments, meeting developmental gates for the ISF, transferring security responsibility to Iraqis). As these major events took months and even years to accomplish, I found that they did not compete with the daily reports of casualties and violence as a means of expressing our progress. While I disagreed with using daily casualty and violence levels as *the* measures of our *strategic* progress (they were measures of the enemy's tactical capacity and *a* measure of our overall progress), in retrospect, I believe that, over time, casualties and violence became the de facto measure of strategic progress in Iraq, and I should have forced a more in-depth discussion with my civilian leadership about their strategic expectations.

I had civil-military interaction with three Iraqi prime ministers and three different sets of cabinet ministers. I treated the Iraqi leaders with the respect due civilian political leaders, and worked to provide them with the key elements of military advice necessary for their decisionmaking. The list I provided Prime Minister al-Maliki when I departed was a compilation of the key areas I had come to believe that civilian and military leaders should discuss in preparing for military operations. As with any difficult issue, I found that productive civil-military interaction is an iterative process that requires a continuous dialogue among civil-military principals until a common understanding is reached. I found that this common understanding is heightened by clearly sharpening differences of opinion rather than papering over them to gain consensus.

In the latter months of 2006 and early 2007, I was consumed with civil-military interaction with civilian leaders in both Baghdad and Washington. As we finalized our plan to secure Baghdad, we worked with Iraqi leadership to cement Iraqi political support for the mission and gain their commitment to the plan's success. The Ambassador and I had long daily sessions with the prime minister and his security ministers, pounding out the details of the plan and ensuring our forces would have freedom of action once they were committed. Once the plan was approved just before Christmas, we turned our attention to the execution of the plan, a phase that required fairly constant interaction with Iraqi leaders that continued through my departure in February.

Simultaneously, we were participating in the Washington review of Iraq policy and strategy that also concluded just before Christmas 2006. The review involved numerous long sessions by video teleconference and had an implementation phase for the announcement and execution of the new policy that continued through January. The transition between Secretary Rumsfeld and Secretary Gates in November and December further complicated the civil-military situation.

Someone told me once that the decisionmaking process at the national level is "idiosyncratic at best." That is an important lesson for future leaders when providing military advice. Do not look for the Military Decision Making Process at the national level. When it comes to providing military advice, yours is only one part of the President's decision calculus. Provide your military advice with your rationale and the courage of your convictions and, as with any military decision, stand by to execute the decision.

Political-Military Integration

Political and military actions must be mutually reinforcing, particularly when operating inside other sovereign countries. With Prime Minister al-Maliki, I used the analogy of the two of us rowing a boat. If I pulled on the "military" oar and he did not pull on the "political" oar, the boat went around in circles. If he pulled on the "political" oar and I did not pull on the "military" oar, the boat went around in circles. If we both pulled together, the boat went forward. I had mixed success with three Iraqi prime ministers in "rowing the boat."

We had our best success integrating political and military actions with the Coalition Provisional Authority–appointed Interim Iraqi Government. We learned early on that Iraqi political support was essential to having the time to bring our military operations to successful conclusion. In the Najaf operations in 2004, careful melding of the political and military efforts yielded the IIG its first success. In the Fallujah operations later that year, actions by the prime minister to disband the Fallujah police force (the terrorists had put on police uniforms) and declare a 24-hour curfew greatly facilitated our tactical operations, and the government's public support of the operation gave us the time we needed to complete the mission. In both of those operations,

INSIGHTS FOR LEADERS

Embassy leadership was kept abreast of the planning, to include participation, with ISF leadership in a rehearsal of the operational concept. In providing security for the January 2005 elections, the imposition of last-minute curfews and driving bans by the interior minister at our request helped disrupt the insurgents' ability to affect the elections. I found that it was not necessary to share tactical details, but giving political and diplomatic leaders a broad idea of what to expect greatly facilitated their ability to support the operation.

Political-military interaction was less productive with both of the elected Iraqi governments that followed. I can only surmise that the greater demand for sovereignty by both subsequent governments affected their ability and willingness to take political risk to support Iraqi and coalition military operations. We had some success with the Iraqi Transitional Government in winning support for the Tal Afar operation in September 2005 and with the agreement the ITG made with Anbar provincial leaders in early 2006 to bring Anbaris into the security forces, to provide money for reconstruction in Anbar, and to release some Anbari prisoners. We were not able to gain their support for weapons and militia bans that would have facilitated our operations to secure Baghdad.

I go into some detail about the political-military integration with the constitutionally elected government of prime minister al-Maliki. The desire of the government for greater say in security actions and a differing view of the threat created frictions that took some months to get through. That said, the Prime Minister's Army Day speech in January of 2007 is a good example of political leaders building public support for military action.

I believe our efforts in 2006 suffered, at least in part, because of the disagreement between the prime minister and the coalition on the nature of the threat and the lack of a political timeline to drive Iraqi actions to resolve their differences over the division of political and economic power—the issues at the heart of the sectarian violence. Whether it was possible to reach agreement on either of these issues at that time is, I believe, an open question. They were decisions for the sovereign government of Iraq, and our government could only attempt to influence them. We could not impose U.S. solutions. The integration of political-military efforts is always difficult, but it is even more so when operating with another sovereign government. It will remain essential to attaining our national objectives in 21st-century conflict.

Momentum and Transitions

In extended campaigns, transitions and their accompanying loss of momentum are inevitable. This was the case in Iraq as we confronted numerous transitions at every level within MNF-I, the U.S. Embassy, and the Iraqi government. Sustaining momentum is not easy, but it is essential to long-term success.

I found that determining whether we had momentum was more art than science. In long operations, things unfold so slowly that it is often hard to tell whether you are moving at all. Our maxim was, "If you're not moving forward, you're moving backward." Leaders need to develop a way to "feel" momentum. A structured assessment process helps, but I found that I got my best sense from my face-to-face meetings with subordinate commanders and Iraqi leaders on their own turf. I learned to judge whether they were comfortable or uncomfortable answering my questions about progress.

Momentum at the theater level generally comes from big events such as successful elections, the passage of major legislation, decisive military victories, and major agreements. The initial UN timeline offered the opportunity in 2005 to sustain momentum through four major events—the initial elections in January 2005, development of the constitution in August, constitutional referendum in October, and elections for the constitutional government in December. To get there, in the absence of political events, we generated momentum through the military successes in Najaf, Samarra, and Fallujah in 2004 and by energizing the development of the Iraqi security forces. Unfortunately, the protracted government formation processes, limited government experience of most of the appointed ministers, and turbulence of three government transitions in 2 years severely limited our ability to sustain political momentum to complement our military efforts.

On the military side, the terrorists and insurgents learned not to mass against us after our successes in Najaf, Fallujah, Tal Afar, and the western Euphrates. So while we maintained momentum and pressure on the terrorists at the tactical level, we did so through daily small unit actions, and it took time for those successes to gain strategic significance. The exception was the killing of Abu Musab al-Zarqawi in June of 2006, a tactical action that had strategic impact.

In an attempt to generate political momentum in 2006, the Ambassador and I developed a series of benchmarks—Iraqi political and security actions that, when taken, would begin to resolve the fundamental tensions over the sharing of political and economic power. By assigning these events a completion date, we hoped to string together a series of political successes that would continue moving the country in a positive direction. By linking these with military operations, we hoped to break the sectarian stalemate that was strangling the county. Unfortunately, the idea never gained the committed support of the newly elected Iraqi leaders.

On the military side, the semiannual and annual transitions of units and staffs affected our momentum, but, largely because of the significant effort made by the Services to prepare their forces, the substantial interaction that took place between units before the new units arrived, and our in-theater training and integration efforts, we were able to somewhat mitigate the impact. I began visiting all newly arrived brigades in early 2005 within 30 days of their arrival to give them a theater overview and to ensure that the leadership clearly understood their mission. With the development of the Phoenix Academy in early 2005 and the COIN Academy in November 2005, I spoke to every class, providing an overview similar to what I provided the brigades. In order to maintain momentum, I felt that it was important incoming leaders heard my expectations directly from me.

I was generally pleased with the unit transition process, but usually I found during my post-transition visits that there was something major that got dropped. For example, the troops that came into an area after a major battle usually did not have the same intensity and commitment to the reconstruction effort as those that had won the victory, and new troops generally seemed to believe that the war began with their arrival. It was human nature at work. The post-transition visits helped with maintaining continuity and momentum.

Maintaining momentum through political and military transitions is another area that is more art than science, and an area of important effort for senior leaders.

Sustaining Yourself

One of the toughest challenges for senior leaders in deployed environments is to sustain their physical, mental, and emotional fitness at levels that allow them to deal with the complex challenges confronting them. I watched four corps's worth of senior leaders come through Iraq. I encouraged each of them to establish a regimen where they got sufficient rest, exercise, and intellectual stimulation so that they could provide their subordinates the direction they needed for success in Iraq. I told them that to sustain themselves for the duration of the mission, they needed to find quality time every day to *REST*: read-exercise-sleep-think. I had found this a useful formula for myself during my time in Bosnia and began to share it with my subordinate leaders as they entered Kosovo in 2000. I practiced it myself in Iraq.

Read. Sometimes the hardest thing to come by after you have been deployed for a while is a fresh idea. Staffs, especially when there are frequent rotations, tend to fall into repeating "facts" based on shared conventional wisdom. I strongly encouraged leaders to find quiet time daily to read something besides their email, their inbox, or intelligence as a way to stimulate new insights. I read every night before I went to sleep and found that it had the added benefit of slowing a mind that was spinning with the events of the day down to the point I could get to sleep. I read a wide variety of books, from T.E. Lawrence's *Seven Pillars of Wisdom* to David

McCullough's *1776* to Stanley Karnow's *Vietnam: A History*. All stimulated useful insights.

Exercise. I strongly encouraged my senior leaders to get on an exercise regimen as soon as they could after their transition process was complete. I made the time to exercise four or five times a week and found it a great way not only to avoid fatigue but also to burn off stress and frustration, of which there was plenty. It was also quiet time alone to think.

Sleep. My experience with U.S. officers and noncommissioned officers is that they tend to push themselves too hard and think that they can get by on less sleep than they really need. In long operations, leaders have to force themselves to get the rest that they need to be most effective. The issues they will be confronted with require them to be at their best.

Think. I found that I needed private time to think, daily and periodically, to keep things straight in my own mind and to be able to shape clear guidance for the staff. I organized my day so that every morning I had 30 minutes to review the intelligence and 30 minutes to think about the previous day and organize my thoughts for the days ahead. Once the day began, there was precious little time for reflection. After a few months on the ground, I began taking a day off every month. I would stay at my quarters, exercise, read, and think about the longer term. Because I found that forcing myself to write things out caused me to sharpen my personal thinking on issues, I would often write something at the end of the day to capture my thinking. After a year, I found that 1 day a month was not enough, and I began taking a half-day off every week. I encouraged my subordinate leaders to do the same.

INSIGHTS FOR LEADERS

Over time, I learned to watch myself to know when I was not at my best. If I got to the point where I did not feel like I was capable of providing creative inputs to the challenges we were dealing with, I looked for the opportunity to get a short break. I also made it a point to take at least a week off outside of Iraq every year and to ensure that all of my subordinates took advantage of R&R leave. Preserving your physical, mental, and emotional strength is critical to the ability to lead at the strategic level.

Operation *Iraqi Freedom* is part of the larger story of the United States of America adapting to the security challenges thrust on us by the al Qaeda attacks of September 11, 2001. The world we live in is in a period of continuous and fundamental change as technology's continuous march ties us closer and closer together and puts the instruments of catastrophic destruction in the hands of nonstate actors. As a result, war in the 21st century will not be like the conventional war that I spent 30 years of a 40-year career training to fight. It will also not be just like Iraq or Afghanistan. At the tactical level, it will be as uncertain and as difficult and as brutal as war has always been. I believe, however, that the complexities of the international security environment will only increase at the operational and strategic levels, bringing greater challenges for senior leaders. We will require agile, adaptive senior leaders to handle the challenges of war in the second decade of the 21st century. It is my hope that this book will contribute to the development of those leaders.

APPENDIX 1: ABBREVIATIONS

A ASSESS: assessments
   AQI: al Qaeda in Iraq

B BG: brigadier general (U.S. Army one-star)

C CAMP PLANS: campaign planning
   CASB: Commander's Assessment and Synchronization Board
   CIA: Central Intelligence Agency
   CIG: Commander's Initiative Group
   CMO: civil-military operations
   COIN: counterinsurgency
   COALITION: coalition operations
   COMMNF-IZ: Commander of Multi-National Force–Iraq
   CONT PLANS: contingency planning
   COR: Council of Representatives
   CORD/SYNC/BRIEF: coordination/synchronization briefing
   COS: chief of staff
   CPIC: Combined Press Information Center
   CPR: Campaign Progress Review
   CSM: command sergeant major

D DCG: Deputy Commanding General
   DCG Detainee Ops: Deputy Commanding General for Detainee Operations
   DCS CIS: Deputy Chief of Staff for Communications and Information Systems Operations
   DCS INTEL: Deputy Chief of Staff for Intelligence Operations

DCS POL/MIL/ECON: Deputy Chief of Staff for Political-Military-Economic Effects

DCS STRAT OPNS: Deputy Chief of Staff for Strategic Operations

DCS STRAT, PLANS & ASSESSMENT: Deputy Chief of Staff for Strategy, Plans, and Assessment

DCS STRATCOM: Deputy Chief of Staff for Strategic Communication

DCS RESOURCES & SUSTAIN: Deputy Chief of Staff for Resources and Sustainment

DEP IRMO: Deputy for Iraq Reconstruction Management Office

DEP STRAT PLANS: Deputy for Strategic Plans

DOS SPT: Department of State support

E   ECON EFF: economic effects

F   FAR: flat-assed rules

G   GEN: general (U.S. Army four-star)

GRD: Gulf Region Division

I   IG: Inspector General

IIG: Iraqi Interim Government

ISF: Iraqi security forces

ISG: Iraq Survey Group

ITG: Iraqi Transitional Government

J   JCC: Joint Contracting Command

JVB: Joint Visitor's Bureau

L   LTG: lieutenant general (U.S. Army three-star)

LOG: logistics

M   MG: major general (U.S. Army two-star)

APPENDIX 1

	MIN COORD: [Iraq] Minister Coordination
	MNB: Multi-National Brigade
	MNC-I: Multi-National Corps–Iraq
	MND: Multi-National Division
	MNF-I: Multi-National Force–Iraq
	MNSTC-I: Multi-National Security Transition Command–Iraq
	MOD: Ministry of Defense
	MOI: Ministry of Interior
N	NSC: National Security Council
	NSPD: National Security Presidential Directive
	NSVI: National Strategy for Victory in Iraq
O	OIF: Operation *Iraqi Freedom*
	OPNS: operations
	OTF: Operation *Together Forward*
P	PAO: public affairs officer
	PERS: personnel
	PLANS: planning
	POLICY DEV/INT: Policy Development and Integration
R	RM: Resource Management
S	SCJS: Secretary of the Combined and Joint Staff
	SJA: Staff Judge Advocate
	SOC: Special Operations Command
	STRATEGY: strategy planning

T	TACON: tactical control
	TF 134: Task Force 134 (Detainee Operations)
	TF 6-26: Task Force 6-26 (Special Operations Task Force)
	TRA: transition readiness assessment
U	UK: United Kingdom
	UN: United Nations
	UNSCR: United Nations Security Council Resolution
	USCENTCOM: U.S. Central Command

APPENDIX 2: THE COALITION, JULY 2004

Albania	Macedonia
Australia	Moldova
Azerbaijan	Mongolia
Bulgaria	Netherlands
Czech Republic	New Zealand
Denmark	Norway
El Salvador	Republic of the Philippines
Estonia	Poland
Georgia	Portugal
Hungary	Romania
Italy	Singapore
Japan	Slovakia
Jordan	Thailand
Kazakhstan	United Arab Emirates
Republic of Korea	Ukraine
Latvia	United Kingdom
Lithuania	

Operation *Iraqi Freedom* (OIF) was not just a seminal experience for the U.S. military. Forces from more than 38 countries contributed effectively to the operation. As with the U.S. forces, the OIF experience had transformative effects on all the militaries that participated. The list above shows the 33 countries that were providing 23,000 forces to OIF when I assumed command in July of 2004. Armenia and Bosnia-Herzegovinia were added in 2005.

Coalition participation held fairly constant throughout my command tenure until completion of the United Nations timeline in December 2005 when it began to decrease. By the end of 2006 we had about two-thirds of the coalition forces that we had when I arrived in 2004.

APPENDIX 3: LEADERSHIP IN IRAQ, 2004–2007

Iraqi Leaders

June 28, 2004–May 3, 2005: Iraqi Interim Government

    Prime Minister: Ayad Allawi

    Minister of Defense: Hazem Shaalan

    Minister of Interior: Falah Hassan al-Naqib

    National Security Advisor: Dr. Mowaffak al-Rubaie*

May 3, 2005–May 20, 2006: Iraqi Transitional Government

    Prime Minister: Ibrahim al-Jafari

    Minister of Defense: Saadoun al-Dulaimi

    Minister of Interior: Bayan Baqir Solagh

    National Security Advisor: Dr. Mowaffak al-Rubaie

May 20, 2006–June 14, 2010: Government of Iraq

    Prime Minister: Nuri al-Maliki

    Minister of Defense: Qadir Obeidi

    Minister of Interior: Jawad al-Bulani

    National Security Advisor: Dr. Mowaffak al-Rubaie

U.S. Ambassadors

    June 2004–March 2005: John D. Negroponte

    March 2005–July 2005: James F. Jeffrey (U.S. Deputy Chief of Mission and U.S. Chargé d'affaires)

    July 2005–March 2007: Zalmay Khalilzad

---

* Dr. Rubaie was appointed to a 5-year term in 2004.

APPENDIX 3

MNF-I Leaders

Commander

GEN George Casey, USA, July 2004–February 2007

Command Sergeant Major

CSM Jeff Mellinger, USA, August 2004–May 2007

Deputy Commanding General

Lt Gen John McColl (UK), May 2004–October 2004

Lt Gen John Kiszley (UK), October 2004–April 2005

Lt Gen Robin Brims (UK), April 2005–October 2005

Lt Gen Nick Houghton (UK), October 2005–February 2006

Lt Gen Rob Fry (UK), March 2006–September 2006

Lt Gen Graham Lamb (UK), September 2006–March 2007

Chief of Staff

MajGen Joe Weber, USMC, March 2004–April 2005

MajGen Tim Donovan, USMC, May 2005–May 2006

MajGen Thomas "Tango" Moore, USMC, May 2006–May 2007

MNC-I Commander

LTG Tom Metz, USA, January 2004–January 2005 (III Corps)

LTG John Vines, USA, January 2005–January 2006 (XVIII Airborne Corps)

LTG Peter Chiarelli, USA, January 2006–December 2006 (V Corps)

LTG Ray Odierno, USA, December 2006–December 2007 (III Corps)

MNSTC-I Commander

LTG Dave Petraeus, USA, June 2004–September 2005

LTG Marty Dempsey, USA, September 2005–March 2007

Special Operations Task Force
    LTG Stan McChrystal, USA, September 2003–August 2008

Deputy Commanding General for Detainee Operations
    MG Geoff Miller, USA, April 2004–October 2004
    MG Bill Brandenburg, USA, November 2004–November 2005
    MG Jack Gardner, USA, November 2005–December 2006
    MajGen Doug Stone, USMC, December 2006–September 2007

Gulf Region Division–Corps of Engineers
    MG Tom Bostick, USA, June 2004–June 2005
    MG Bill McCoy, USA, June 2005–October 2006
    BG Mike Walsh, USA, October 2006–October 2007

Joint Contracting Command
    MG John Urias, USA, January 2005–January 2006
    Maj Gen Darryl Scott, USAF, February 2006–October 2007

Deputy Chief of Staff for Strategic Effects
    MG Hank Stratman, USA, June 2004–July 2005
    MG Rick Lynch, USA, June 2005–July 2006
    MG Bill Caldwell, USA, June 2006–May 2007

Deputy Chief of Staff for Strategy, Plans, and Assessment
    Maj Gen Steve Sargent, USAF, December 2003–May 2005
    Maj Gen Rusty Findley, USAF, May 2005–May 2006
    Maj Gen Kurt Cichowski, USAF, May 2006–May 2007

Deputy Chief of Staff for Strategic Operations
    MG Tom Miller, USA, July 2003–August 2004
    MG Jim Molan (AUS), September 2004–April 2005
    MG Eldon Bargewell, USA, April 2005–June 2006
    MG Dave Fastabend, USA, June 2006–June 2007

Joint Interagency Task Force–High Value Individuals

BG Frank Kearney, USA, August 2004–July 2005

BG Craig Broadwater, USA, August 2005–January 2006

Brig Gen Mike Longoria, USAF, January 2006–July 2006

Deputy Chief of Staff for Intelligence

MG Barbara Fast, USA, July 2003–August 2004

MG John Defreitas, USA, August 2004–July 2005

MG Rick Zahner, USA, July 2005–October 2006

BG Dave Lacquement, USA, October 2006–September 2007

Deputy Chief of Staff for Coalition Operations

BG De Pascale (HUN), May 2004–November 2004

BG Alessio Cecchetti (IT), October 2005–April 2006

BG Pier Paolo Lunelli (IT), May 2006–November 2006

BG Dan Neagoe (ROM), November 2006–February 2007

Deputy Chief of Staff for Communications and Information Systems

RADM Nancy Brown, USN, August 2004–March 2005

Brig Gen Rick Dinkins, USAF, March 2005–December 2005

Brig Gen Gary Connor, USAF, December 2005–December 2006

Brig Gen Ronnie Hawkins, USAF, December 2006–December 2007

Deputy Chief of Staff for Resources and Sustainment

BG Scott West, USA, July 2003–June 2004

BG Gerry Minetti, USA, July 2004–July 2005

MG Kathy Gainey, USA, July 2005–September 2006

BG Steve Anderson, USA, September 2006–August 2007

APPENDIX 4: IRAQI SECURITY FORCES PROGRESSION, 2004–2007

The training and equipping of the Iraqi security forces was a significant accomplishment, especially because the ISF were organized, trained, equipped, and then put directly into combat operations with our forces. The transition team and partnership programs enabled us not only to complete the training and equipping of the ISF in an active combat environment, but also to instill in them the qualities of a professional military operating under civilian leadership.

The quantitative growth of the ISF is shown below, from approximately 90,000 trained and equipped military, police, and border forces in July 2004 to over 325,000 by the end of January 2007, a growth of over 3.5 times.

The qualitative growth of the Iraqi army was equally dramatic. In June 2005, we began monthly reporting on army and national police units to track their progress. There was a four-fold increase in army battalions conducting operations independently or with coalition force support (TRA 1 and 2) between June 2005 and January 2007 (24 in June 2005, 96 in January 2007)—a period of only 19 months.

The improvement in national police battalions was not as dramatic, due largely to their infiltration by sectarian influences in the summer and fall of 2005, which required the retraining of all national police units. Each brigade was pulled offline, its leadership purged of sectarian influence, and then retrained.

APPENDIX 4

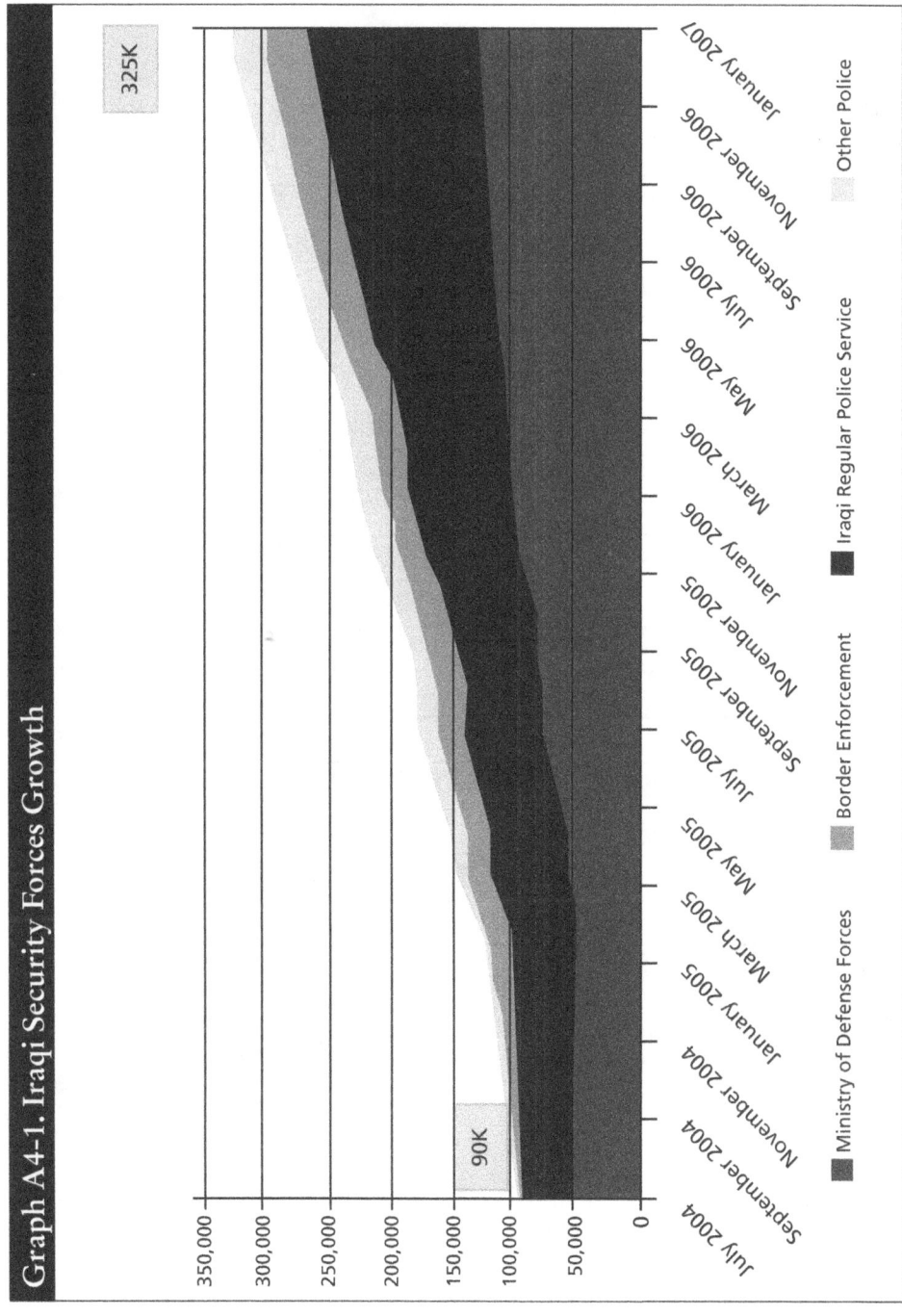

Graph A4-1. Iraqi Security Forces Growth

STRATEGIC REFLECTIONS

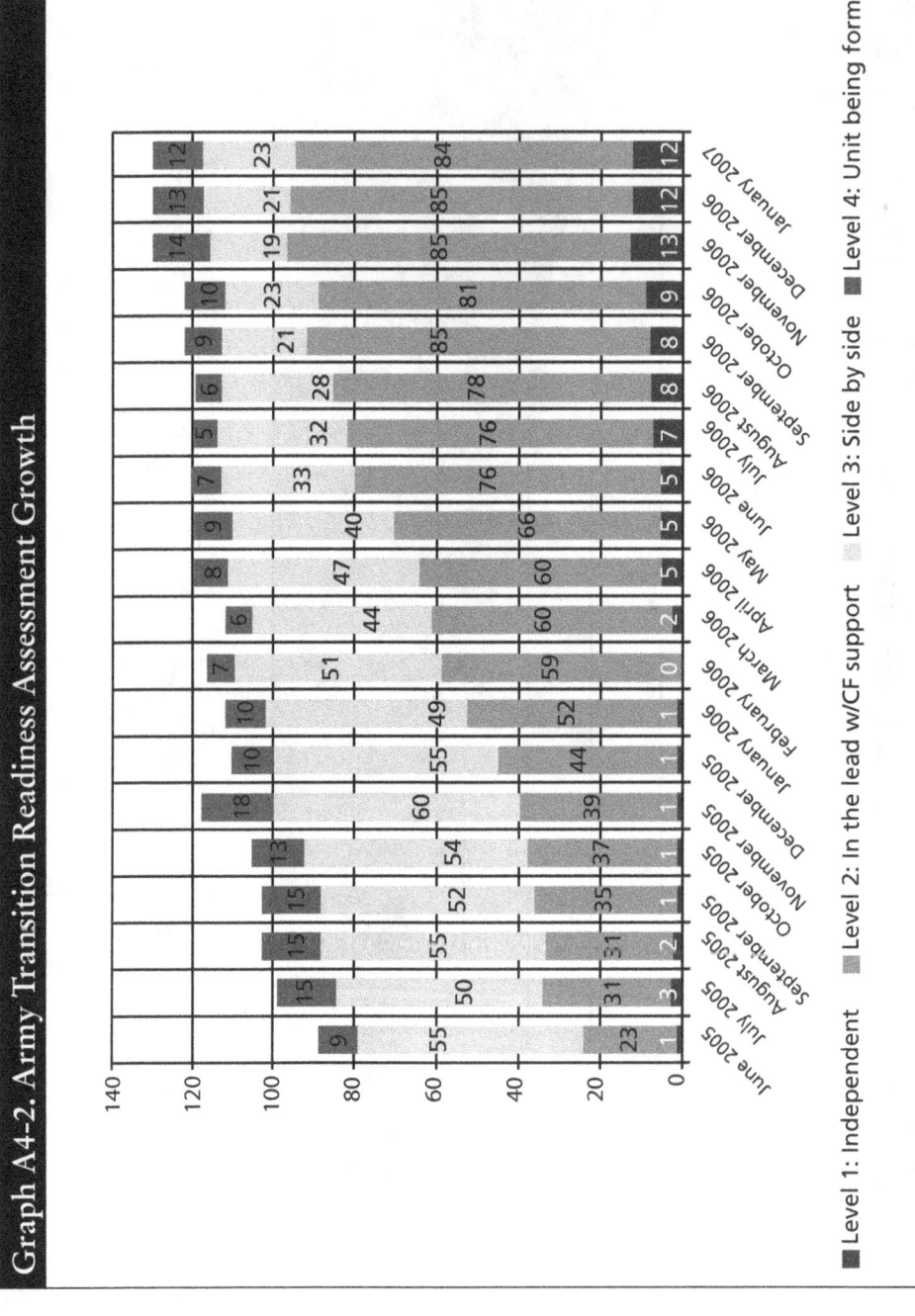

Graph A4-2. Army Transition Readiness Assessment Growth

190

APPENDIX 4

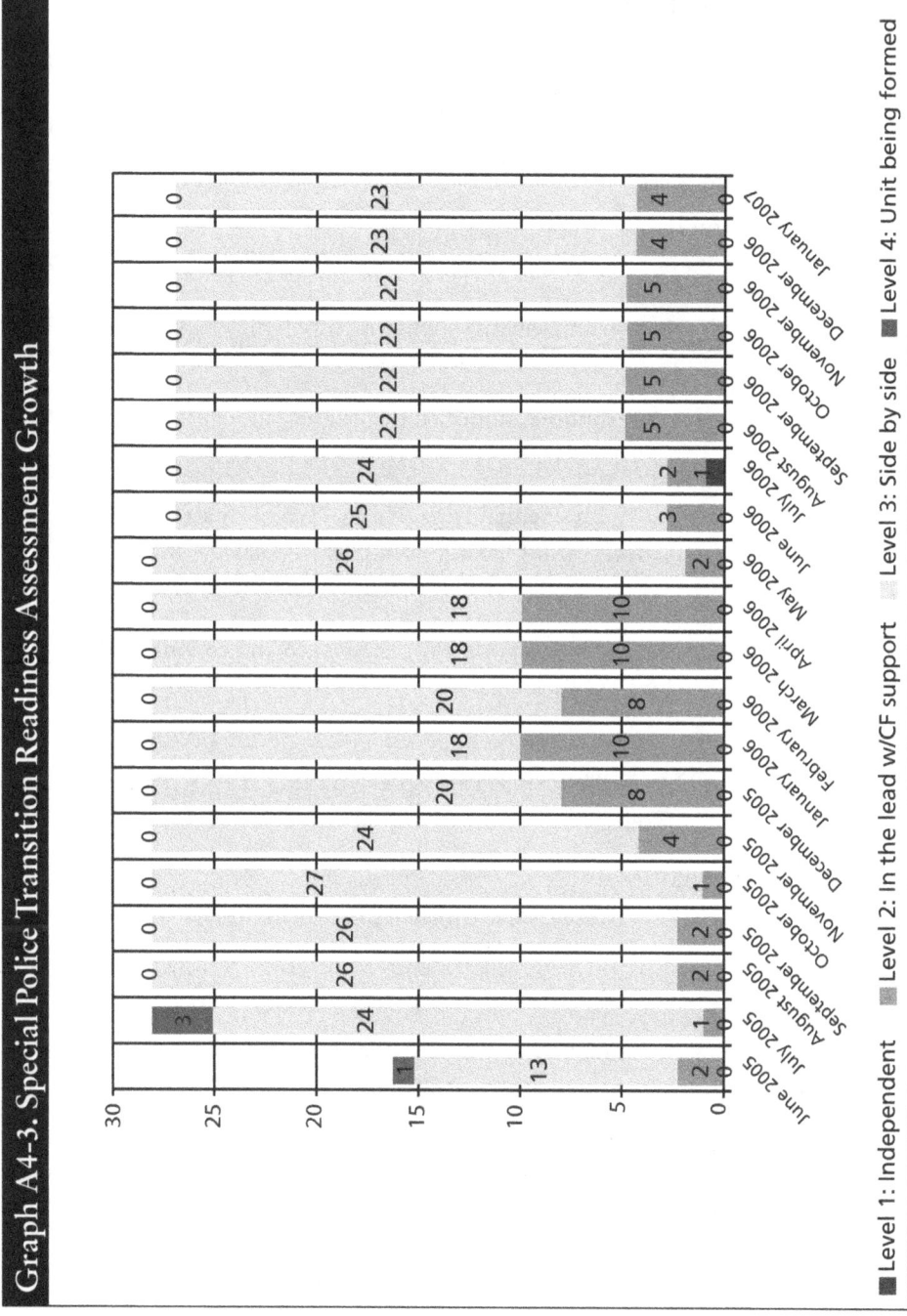

APPENDIX 5: OPERATION *IRAQI FREEDOM* CHRONOLOGY, 2004–2007

March 19, 2003
Operation *Iraqi Freedom* begins.

April 19, 2004
John Negroponte nominated as Ambassador to Iraq.

May 6, 2004
Ambassador Negroponte confirmed by Senate.

May 11, 2004
National Security Presidential Directive 36, "United States Government Operations in Iraq," issued.

May 15, 2004
Multi-National Force–Iraq (MNF-I) established and commanded by LTG Rick Sanchez.

May 24, 2004
President George W. Bush delivers speech on Iraq at Army War College.

May 26, 2004
GEN George W. Casey, Jr., nominated to command MNF-I.

June 8, 2004
United Nations Security Council Resolution (UNSCR) 1546 unanimously approved by Security Council.

June 14–20, 2004
GEN Casey visits Iraq with Deputy Secretary of Defense Paul Wolfowitz.

June 26, 2004
GEN Casey confirmed by Senate to command MNF-I.

APPENDIX 5

June 28, 2004
U.S. Government transfers sovereignty to Iraq.

July 1, 2004
GEN Casey takes command of MNF-I.

July 7, 2004
First meeting of Strategic Action Committee, a deputies-level meeting to frame security issues for Iraqi and U.S. leaders.

July 8, 2004
First meeting of Ministerial Committee for National Security, the primary forum for Iraqi and U.S. leaders to jointly address security issues.

July 15, 2004
Red Team assessment, "Building Legitimacy and Confronting Insurgency in Iraq," published. Assessment completed at joint request of Ambassador Negroponte and GEN Casey.

August 5, 2004
MNF-I 2005 campaign plan, "Operation Iraqi Freedom—Partnership: From Occupation to Constitutional Elections," published.

August 5–27, 2004
Battle of Najaf.

August 18, 2004
U.S. Embassy/MNF-I joint mission statement published.

September 2004
MNF-I counterinsurgency study conducted as historical review of best practices.

November 2, 2004
President Bush is reelected.

November 8–18, 2004
Battle of Fallujah.

December 5, 2004
First Campaign Progress Review published.

December 14–20, 2004
GEN Casey returns to Washington for consultations. Also visits Fort Bragg to direct incoming XVIII Corps to begin preparing to implement partnership and transition teams on arrival with internal assets.

January 30, 2005
Iraqis vote to elect a Transitional National Assembly in first democratic elections in Iraq since 1954. Eight million Iraqis—58 percent of electorate—turn out to vote.

February–May 2005
Iraqi government transitions from Iraqi Interim Government (IIG) to Iraqi Transitional Government (ITG).

February 7, 2005
Second U.S. Embassy/MNF-I joint mission statement, "A Plan for the Year Ahead: Transition to Self-Reliance," published.

February 10, 2005
Multi-National Corps–Iraq (MNC-I) transition of authority from III Corps, commanded by LTG Tom Metz, to XVIII Airborne Corps, commanded by LTG John Vines.

March 2005
Ambassador Negroponte departs Iraq.

April 2005
Phoenix Academy, a training center for incoming transition teams, established.

April 22, 2005
MNF-I campaign action plan, "Transition to Self-Reliance," published, emphasizing development of Iraqi security forces (ISF) capacity and establishing transition team and partnership programs and initial framework for transition of security responsibility to Iraqis.

May 2005
Transition readiness assessment developed to measure ISF capabilities. First report delivered in June.

May 2005
ITG seated. Transitional National Assembly begins drafting constitution.

June 15, 2005
Seven Provincial Support Teams, the predecessor to Provincial Reconstruction Teams (PRTs), established.

June 22, 2005
Ambassador Zalmay Khalilzad arrives.

June 22–28, 2005
GEN Casey visits Washington for consultations.

June 27, 2005
Second Campaign Progress Review (December 2004–June 2005) published.

July–August 2005
MNF-I conducts survey to determine how well coalition forces are applying counterinsurgency doctrine.

July–December 2005
MNF-I military operations focus on restoring Iraqi control to Iraq's western border. Third Armored Cavalry Regiment reinforces Tal Afar–Mosul corridor. Western Euphrates River Valley campaign conducted in Anbar Province.

August 16, 2005
U.S.-Iraqi Joint Committee to Transfer Security Responsibility established to set conditions for gradual transition of security to appropriate Iraqi authorities.

September 2005
Intelligence task force established to track sectarian violence in Iraq.

September 2005
COIN Academy established; first class conducted in November 2005.

September 5, 2005
Command of Multi-National Security Transition Command–Iraq changes from LTG David Petraeus to LTG Martin Dempsey.

September 10, 2005
Assessment on transition teams published.

September 27–October 3, 2005
GEN Casey visits Washington for consultations.

October 2005
Responsibility for developing Ministry of the Interior (MOI) moved from U.S. Embassy to MNF-I.

October 15, 2005
Iraqis approve constitution drafted by Transitional National Assembly. Ten million Iraqis vote; 78.6 favor constitution.

October 30, 2005
"Strategic Planning Directive (November 2005–April 2006)—Making the Elections Decisive" published to bridge uncertainty of new government formation period. Tenets include "Al Qaeda out," "Sunni in," and "ISF in the lead."

November 2005
PRTs approved for Mosul, Kirkuk, and Babil provinces.

November 30, 2005
Illegal MOI detention facility discovered in Baghdad.

November 30, 2005
National Strategy for Victory in Iraq issued by White House.

December 6, 2005
Third U.S. Embassy/MNF-I joint mission statement published.

December 15, 2005
11.8 million Iraqis (76 percent of registered voters) elect members of Iraqi assembly under new constitution; government formation period begins.

December 20, 2005
Third Campaign Progress Review published.

December 24, 2005
First off-ramp of two coalition brigades announced following successful completion of UNSCR 1546 political timeline.

January 19, 2006
MNC-I transition of authority from XVIII Airborne Corps to V Corps, commanded by LTG Peter Chiarelli.

February 22, 2006
Al-Askari Mosque in Samarra bombed, inflaming sectarian tensions.

February 24, 2006
MNF-I publishes "Strategic Directive: Golden Mosque Bombing" to direct actions in aftermath of al-Askari Mosque bombing.

March 14–June 14, 2006
Operation *Scales of Justice* stabilizes situation in Baghdad sufficiently to allow establishment of constitutionally elected government of Iraq.

April 21, 2006
Nuri al-Maliki chosen to replace Prime Minister Ibrahim al-Jafari, becoming the first democratically elected prime minister of Iraq under new constitution.

April 28, 2006
New joint campaign plan, "Operation Iraqi Freedom Transition to Iraqi Self-Reliance," published, projecting December 2009 as time when Iraqis would be self-reliant for security.

May 20, 2006
Prime Minister al-Maliki's government confirmed by Iraqi parliament. Security ministers not confirmed for 2 more weeks.

June 7, 2006
Abu Musab al-Zarqawi, the leader of al Qaeda in Iraq, killed during U.S. air raid.

June 7, 2006
New security ministers confirmed.

June 9, 2006
Joint campaign action plan, "Unity, Security, Prosperity," published.

June 12, 2006
Camp David discussions. Ambassador Khalilzad and GEN Casey participate by secure video teleconference.

June 14, 2006
Fourth Campaign Progress Review published.

APPENDIX 5

June 15–August 6, 2006
Operation *Together Forward I* (Baghdad security), the new government's first attempt to halt sectarian violence in capital, meets with initial success but flounders as ISF do not consistently hold cleared areas.

June 19–23, 2006
GEN Casey visits Washington for consultations.

July 13, 2006
Muthanna becomes first Iraqi province to assume security responsibility.

July 18, 2006
Anticipated off-ramp of U.S. forces canceled.

July 19, 2006
Joint Committee to Achieve Iraqi Security Self-Reliance established to refine framework for Iraqi assumption of security responsibility, continuing work of Joint Committee to Transfer Security Responsibility.

July 28, 2006
172nd Stryker Brigade extended 4 months in Iraq to address sharp increase in sectarian violence.

August 7–October 22, 2006
Operation *Together Forward II* (Baghdad security) initiated to reduce sectarian violence before Ramadan.

September 21, 2006
Dhi Qar becomes second Iraqi province to assume security responsibility.

October–December 2006
U.S. Government review of Iraq policy.

October 7–11, 2006
GEN Casey visits Washington for consultations.

November 7, 2006
U.S. midterm elections.

November 8, 2006
Donald Rumsfeld resigns as Secretary of Defense and President Bush nominates Robert Gates.

November 30, 2006
Meeting in Amman, Jordan, with President Bush and Prime Minister al-Maliki.

December 6, 2006
Iraq Study Group report released.

December 14, 2006
MNC-I transition of authority from V Corps to III Corps, commanded by LTG Ray Odierno.

December 20, 2006
Najaf becomes third Iraqi province to assume security responsibility.

December 23, 2006
Fifth Campaign Progress Review published, noting that because of sectarian violence, objectives were not being achieved within planned timeframes. Review stated that improvements in bringing all elements of national power to bear were necessary.

December 23, 2006
Prime Minister al-Maliki approves new Baghdad security plan and additional deployment of U.S. forces.

December 30, 2006
Saddam Hussein executed by hanging.

APPENDIX 5

January 6, 2007
In Iraqi Army Day speech, Prime Minister al-Maliki announces government support for ISF and new Baghdad security plan and agrees to take action against "all who break the law."

January 10, 2007
President Bush delivers speech to Nation, announcing deployment of five brigades to Iraq.

February 8, 2007
GEN Casey confirmed as Army chief of staff.

February 10, 2007
GEN Casey passes command of MNF-I to GEN Petraeus.

NOTES

Chapter 1

[1] National Security Presidential Directive 36, "United States Government Operations in Iraq," May 11, 2004.

[2] Ibid.

[3] George W. Bush, speech at the U.S. Army War College, Carlisle, PA, May 24, 2004.

[4] Ibid.

[5] United Nations Security Council Resolution 1546, June 8, 2004.

[6] Ibid.

[7] Ibid.

[8] Donald Rumsfeld, memo to General John Abizaid, "Support for Other Government Activities in Iraq," April 27, 2004.

[9] Donald Rumsfeld, memo, "Some Thoughts on Iraq and How to Think about It," June 7, 2004.

[10] Senate Armed Services Committee, "Advance Questions for General George W. Casey, Jr., U.S. Army, Nominee for Commander, Multinational Force–Iraq," June 24, 2004.

[11] Bush, speech.

Chapter 2

[1] John P. Abizaid, memo to General Casey, "Initial Guidance," July 1, 2004.

[2] Red Team, "Building Legitimacy and Confronting Insurgency in Iraq," July 15, 2004.

[3] Ibid.

[4] Multi-National Force–Iraq (MNF-I) and U.S. Mission–Iraq, "MNF-I–Embassy Joint Mission Statement," August 18, 2004.

[5] MNF-I campaign plan, "Operation Iraqi Freedom–Partnership: From Occupation to Constitutional Elections," August 5, 2004.

6. Multi-National Security Transition Command–Iraq, presentation to the Secretary of Defense, "IPR [In-Progress Review] with Secretary of Defense," August 4, 2004.

7. Multi-National Security Transition Command–Iraq, presentation to the Secretary of Defense, "IPR with Secretary of Defense," July 14, 2004.

8. Bill Hix and Kalev Sepp, "Successful and Unsuccessful Counterinsurgency Practices," January 22, 2005.

9. MNF-I, "MNF-I 5-Month Campaign Progress Review," December 5, 2004.

10. George W. Casey, Jr., "Iraq Update," December 16, 2004.

Chapter 3

1. George W. Casey, Jr., memo to General Abizaid, "The Way Ahead for 2005," January 5, 2005.

2. Ibid.

3. MNF-I, "Joint Mission Statement," February 7, 2005. Emphasis added.

4. Ibid.

5. MNF-I, "Joint Campaign Action Plan," April 22, 2005.

6. MNF-I, "Campaign Progress Review (December 2004–June 2005)," June 27, 2005.

7. George W. Casey, Jr., congressional testimonies before the Senate and House Armed Services Committees, June 23, 2005.

8. MNF-I, "Strategic Planning Directive (November 2005–April 2006)—Making the Elections Decisive," October 30, 2005.

9. MNF-I, "The September Assessment [of Transition Teams]," September 10, 2005.

10. George W. Bush, radio address, October 15, 2005.

11. George W. Casey, Jr., presentation to Stephen Hadley, "Iraq Campaign Plan," September 17, 2005.

12. NSC, "National Strategy for Victory in Iraq," November 2005.

13. MNF-I and U.S. Mission–Iraq, "Joint Mission Statement," December 6, 2005.

14. MNF-I, "Campaign Progress Review," December 20, 2005.

Chapter 4

[1] George W. Casey, Jr., presentation to National Security Council, "MNF-I Update," January 5, 2006.

[2] MNF-I, "Strategic Directive: Golden Mosque Bombing," February 24, 2006.

[3] MNF-I and U.S. Mission–Iraq, "Joint Campaign Plan," April 28, 2006.

[4] Ibid.

[5] George W. Casey, Jr., presentation to NSC at Camp David, "Joint Campaign Plan," June 12, 2006.

[6] MNF-I and U.S. Mission–Iraq, "Joint Campaign Progress Review," June 14, 2006.

Chapter 5

[1] George W. Casey, Jr., email to Generals Abizaid and Peter Pace, July 18, 2006.

[2] James A. Baker III and Lee H. Hamilton, "The Iraq Study Group Report," December 2006.

[3] George W. Bush, speech to Nation, January 10, 2007.

Len Rubenstein Photography

ABOUT THE AUTHOR

GEORGE WILLIAM CASEY, JR.

Chief of Staff
United States Army
April 2007–April 2011
Commander, Multi-National Force–Iraq
July 2004–February 2007

General George W. Casey, Jr., was born in Sendai, Japan, on July 22, 1948, into a military family. He was commissioned a Second Lieutenant of Infantry from the Georgetown University School of Foreign Service in 1970. His father, Major General George William Casey, was killed in Vietnam that same year while commanding the 1st Cavalry Division.

Throughout his career, General Casey served in operational assignments in Germany, Italy, Egypt, Southwest Asia, and the United

States. He commanded at every level from platoon to division, including a mechanized infantry battalion at Fort Carson, Colorado, and a mechanized infantry brigade at Fort Hood, Texas. He also served as Assistant Division Commander for Maneuver and Support in the 1st Armored Division in Germany and Bosnia (as part of Operation *Joint Endeavor*) and commanded the 1st Armored Division in Bad Kreuznach, Germany.

His principal staff assignments included Chief of Staff, 1st Cavalry Division, Fort Hood, Texas; Operations Officer and Chief of Staff, V (US/GE) Corps, Heidelberg, Germany; Deputy Director for Politico-Military Affairs, Joint Staff; Commander, Joint Warfighting Center/J7, U.S. Joint Forces Command; Director Strategic Plans and Policy, Joint Staff; Director of the Joint Staff; and 30th Vice Chief of Staff, U.S. Army.

From July 2004 until February 2007, General Casey was the Commander of Multi-National Force–Iraq, a coalition of more than 30 countries, in Operation *Iraqi Freedom*. He led coalition forces through three transitions of sovereign Iraqi governments while increasing the size and capacity of Iraqi security forces and setting the conditions for long-term success.

From April 10, 2007, until April 11, 2011, General Casey served as the 36th Chief of Staff, U.S. Army. Taking over at a time when the Army was stretched from 6 years of continuous combat, General Casey stabilized and transformed the Army into an agile fighting force prepared for the challenges of the 21st century while meeting the demands of two wars.

General Casey holds a Master of Arts Degree in International Relations from Denver University and served as a Senior Fellow at the Atlantic Council of the United States.

INDEX

Abizaid, John, GEN USA, 2, 17, 19, 51, 64, 68, 114, 141, 144, 166
Abud Qanbar, Lt Gen ISF, 134
adaptation, importance of, 159–162
advisors, 57–58
al-Askari Mosque bombing, 89–96. *See also* Samarra bombing
al-Jafari, Ibrahim, Prime Minister of Iraq, 2, 55, 64, 75, 84, 89
al-Maliki, Nuri, Prime Minister of Iraq, 3, 104, 106, 107, 112, 121, 124, 125, 131, 133, 134, 149, 168, 170
   building relationships, 98–102
   guidance to military leaders, 150–151
   meetings with President Bush, 106, 138–139, 146–148
al-Masri, Abu Ayyub, 119
al Qaeda
   after al-Zarqawi's death, 108
   disruption of activities, 84–85
   exhortation to violence, 119
   operations to defeat, 71
   success against, in Anbar, 125
   suicide attacks, 110
al-Sadr, Muqtada, 40
al-Zarqawi, Abu Musab, 79, 103–104
Albu Mahal tribe, 85
Allawi, Ayad, Prime Minister of Iraq, 2, 6, 16, 39, 42, 50, 55, 58, 158
   confidence boost after Najaf, 41
   early meetings, 36–37
   on security challenges, 14–15
Ambassador, United States to Iraq, 7. *See also* Negroponte, John; Jeffrey, James F.; Khalilzad, Zalmay
amnesty agreements, 126
Anbar Province
   cooperation in, 171
   as focus of insurgency, 96, 111, 119, 140
   operations to defeat al Qaeda in, 84–85
   success in, 125
Army Canal, 112
assessment
   in campaign plans, 156
   importance of, 159–162
   tools for, 32–33, 160

Ba'athists
   concerns over return, 92
   as greatest danger, 128
   relative importance, 112–113
Badr Corps, 145
Baghdad
   as center of conflict, 94
   civilian casualties, 140
   as focus of insurgency, 96, 110–111, 119, 140
   Iraqi control of, 131–135
   military operations in, 126–135
   Red Team recommendations, 117
   security assessment by MNC-I, 129
   security concerns, 102
   security efforts redefined, 111
   surge troops deployed, 148–149
   violence against civilians, 110
Balad Airbase, 36
Bridging Strategy, 68, 84–89
Bush, George W., President, 2, 5, 8, 18, 33, 50, 74, 104, 122, 125
   meetings with Prime Minister al-Maliki, 106, 138–139, 146–148
   speech at U.S. Army War College, 8

Camp David consultations, 104–106
Camp Victory, 34
campaign action plans
   2005, 61–62
   rationale for, 156
campaign planning
   after Samarra bombing, 96, 98

INDEX

initial plan 2004, 28–31
intent, 156
joint plan 2006, 96–98
presented to President Bush, 33, 104–105
Campaign Progress Reviews
   2004 December, 46–48
   2005 June, 64–66
   2005 December, 78–79
   2006 June, 107
   2006 December, 200
   described, 160–161
"can-do" attitude, impact on mission, 13, 57, 163
checkpoints, 112, 122–123
Chiarelli, Peter, LTG USA, 3, 88
civil authority
   integrating military efforts, 127
   interactions with military, 165–170
civil war, 94–95
civilians
   attacks on, Baghdad and Diyala, 96, 110–111, 140
   casualties in Baghdad, 140
   as targets of attack, 70
clarity of thought, 154–156
clear-hold-build concept, 41
COIN. *See* counterinsurgency
COIN Academy, 73, 88, 164, 174
command, lines of, 7–8
Commander's Assessment and Synchronization Board, 32–33, 160
commanders' conferences, 161
Commander's Emergency Response Program, 69
communication with civil authorities, 166
communication mission, 31
confirmation hearings
   for chief of staff of the Army, 149
   for commander, MNF-I, 16–17

congressional engagement, 6, 64–66, 135
construction projects, 41
Council of Representatives, 128
counterinsurgency
   academy, 73, 88, 164, 174
   best practices summary, 45, 163
   doctrine application, review, 73, 164
   duration of, 45, 53
   increased presence leads to reduced casualties, 96
   manual for, 164
   "Najaf model," 41–42
   need for popular support, 28
criminals, 27
culture
   importance of understanding Iraqi, 63
   institutional change in, 162–163
   interagency differences in, 158
   organizational, 162–165
curfews, 90

death squads
   attacks by, 110–111
   casualties from, 140
   Iranian training of, 116
   relative importance of, 112–113
decisionmaking process, 169–170
Dempsey, Martin E., LTG USA, 3
Desert Protectors, 85
Diyala Province, 96, 110–111, 140
drawdowns
   2006 cancellation, 113–115
   2006 plans, 83, 107

economic development mission, 30–31
Eisenhower, Dwight D., 3
elections
   2005 January, 49–50
   2005 December, 80
   2005 outcome possibilities, 46
   deadlines for, 9
   planning and preparation, 42–44

INDEX

polarizing effect, 82
security efforts, 43
security planning, 39
suicide attacks as threat, 70–71
Sunni participation, 85–86
exercise, importance for leaders, 176
expectations, setting, 82–83, 166–167

Fallujah
as haven for insurgents, 15
importance of reclaiming, 42
success in, 170
fighter networks, 70
"flat-assed rules," 62, 63
fusion centers, 71

Gates, Robert M., U.S. Secretary of Defense, 2, 144, 146, 148, 169
governance mission, 30–31
Green Zone, 36
group think, avoiding, 162

Hadley, Stephen J., National Security Advisor, 2, 76, 138
headquarters organization, 33–34, 35

IEDs (improvised explosive devices)
cause of casualties, 119
counter-efforts, 88–89
Iranian supply of, 116
Independent Iraqi Election Commission, 43
"Insurgency," as term less useful, 104
insurgency, Red Team view of, 24–25, 27
intelligence updates, 161
Interim Iraqi Government (IIG)
early assessment, 25
formation, 9
Iran
concerns over influence, 92, 93, 95
countering influence of, 116
operatives captured, 145

Iraq Civil Defense Corps. *See also* Iraqi National Guard
inadequacy of, 15
Iraq Study Group, 135–137
Iraq Survey Group, 36
Iraqi army. *See* Iraqi security forces (ISF)
Iraqi government (interim). *See* Interim Iraqi Government (IIG)
Iraqi government (sovereign)
coordination of efforts with, 158–159
differing view of threats, 112–113, 122–123
early establishment of, 19
legitimacy of, 28–29
Iraqi government (transitional). *See* Iraqi Transitional Government (ITG)
Iraqi National Guard, 38
Iraqi people
independent action, encouragement of, 13
view of coalition forces, 54
Iraqi police. *See also* Iraqi security forces (ISF)
2005 updated plans for, 38
collaboration with militia, 119
deficiencies in, 15, 87–88, 110
delayed development, 72
sectarian influence, 91
transfer of responsibility for, 72
"Year of Police," 87–88
Iraqi security forces (ISF)
2005 development priorities, 32
"face of" 2005 elections, 49–50
growth in capability, 105
growth in numbers, 189–191
"ISF in the lead" concept, 87
improving readiness, 54
initial review of, 38
Iraqi vision for, 15–16
transition concept, 60–61
troop goals, 9

211

INDEX

troop levels (mid January 2006), 87
Iraqi Transitional Government (ITG)
  cooperation with, 171
  formation, 9
  transition of, 55–56
Islamic extremists, 27. See also Shia extremists; Sunni extremists
Islamic Revolutionary Guard Corps–Quds Force, 116

Jeffrey, James F., U.S. Ambassador to Iraq, 2
Joint Chiefs of Staff, 12, 48, 64, 107, 124, 137–138, 140
Joint Committee to Transfer Security Responsibility, 75
Joint Contracting Command, 36
joint mission statements
  August 2004, 26
  February 2005, 56–57
  December 2005, 77–78
  intent, 155

Khalilzad, Zalmay M., U.S. Ambassador to Iraq, 2, 69, 75, 82, 85
  relationship with, 67
kidnapping
  and murder of Iraqi, 101–102
  of coalition soldier in Baghdad, 122–123

Lawrence, T.E. (Thomas Edward), 51
leader sustainment, personal, 175–177

McChrystal, Stanley A., LTG USA, 3, 103
Metz, Thomas F., LTG USA, 3
Ministerial Committee for National Security, 16, 37
mission
  changes in after Samarra bombing, 92
  consultations about initial direction, 12–13

duration discussions, 48–49, 53
initial concept development with Negroponte, 11
mission statements, 29, 62, 98, 163
momentum, during transitions, 54, 67–68, 172–174
Multi-National Brigade (MNB) area of operation, 21, 23
Multi-National Corps–Iraq (MNC-I), 30, 32, 54, 63, 67, 70, 84, 111, 112, 129, 144, 159
  responsibilities, 36
  transitions, 88
Multi-National Division (MND) areas of operation, 21, 23
Multi-National Force–Iraq (MNF-I)
  composition in 2004, 21
  establishment, 22
  headquarters organization, 33–34, 35
Multi-National Security Transition Command–Iraq (MNSTC-I), 32, 36, 47
Muthanna Province, 76, 115
Myers, Richard B., GEN USAF, 16th Chairman of the Joint Chiefs of Staff, 2, 13, 18

Najaf
  as haven for insurgents, 15
  reclaiming of, 40
  success in, 170
National Strategy for Victory in Iraq (NSVI), 77
National Security Advisor, 2. See also Hadley, Stephen; Rice, Condoleezza
National Security Presidential Directive, 6–9
Negroponte, John, U.S. Ambassador to Iraq, 18, 19, 20, 39, 53, 54, 77, 154, 157
  relationship with Casey, 10

Ninewah Province, 96, 111

Odierno, Raymond, T., LTG USA, 3
"off-ramp" plans. *See* drawdowns
One Team/One Mission concept, 22, 30, 33, 37, 67, 152
   building and sustaining, 158
   formation, 10–11
   unity of effort, 157
Operation *Iraqi Freedom* chronology, 192–201
Operation *Scales of Justice*, 90–91
Operation *Together Forward*, 110
Operation *Together Forward II*, 118–119

Pace, Peter, GEN USMC, 17th Chairman of the Joint Chiefs of Staff, 2, 114, 135, 137, 138, 140, 144, 145
Partnership concept, 29, 58–59
Petraeus, David H., LTG USA, 3, 20, 146
   change of command, 149, 151
   ISF assessment leadership, 38
Phoenix Academy, 61, 164
policy review, Washington 2006, 135–138
political-military integration, 127, 170–172
Powell, Colin, U.S. Secretary of State, 2, 6, 16
presumption of confirmation, 14
prisons, Iraqi, 79
progress, measuring, 32–33, 46–47, 72, 160, 167–168
Provincial Reconstruction Teams (PRTs), 68–69
public services, 112, 119
public support, as center of gravity, 28

Ramadan, violence during, 118–119
Ramadi, 125
reading, importance for leaders, 175
reconciliation, importance of, 79, 86, 108, 114–115, 125–127, 135–142
reconstruction funding to PRTs, 69

Red Teams
   2006 actions, 116
   avoiding group think, 162
   election outcomes, 46
   initial, 22–28
   subsequent formations of, 67
   unity of effort, 157–158
reporting, balance in, 53
reversing decisions, 115–116
Rice, Condoleezza, U.S. Secretary of State, National Security Advisor, 2
ROC drill, election, 50
rowboat analogy, 150
Rumsfeld, Donald H., U.S. Secretary of Defense, 2, 144, 169
   consultations with, 48, 74, 76, 114
   initial guidance, 13, 18
   resignation, 126, 135
   teleconferences with, 37–38

Sadr City
   car bombs in, 129
   clearance efforts, 112–113
safe haven elimination
   Fallujah, 42
   Najaf, 40
Saladin Province, 140
Samarra bombing, 89–96
Sanchez, Ricardo, LTG USA, 20, 22
Schoomaker, Peter J., GEN USA, 5, 146
Secretary of Defense, U.S., 2. *See also* Gates, Robert M.; Rumsfeld, Donald H.
Secretary of State, U.S., 7. *See also* Powell, Colin; Rice, Condoleezza
sectarian violence
   2005 increase in, 79
   after Samarra bombing, 89–91
   civilian focus, 110–111
   geographic focus, 140
security challenges, Iraqi view of initial, 14–15

INDEX

security mission, 30, 32
Self-Reliance Phase (of Operation *Iraqi Freedom*), 29
sensitive offensive operations, 16
services, 112, 119
Shia extremists
   attacks by, 110–111
   death squads, 140
   defeat in Najaf, 40
   kidnapping by, 101–102
   security concerns, 93, 95
   threat assessment in 2004, 27
Shia Iraqis
   civilians as targets of attack, 70
   inexperience in governing, 53
sleep, importance for leaders, 176
Special Forces, training conventional forces, 61, 164
special operations task force, 36, 70, 84, 103, 113, 117
status updates. *See* Campaign Progress Reviews
Strategic Action Committee, 16, 37
Stryker brigade extension, 115
suicide attacks
   after al-Zarqawi's death, 108
   contribution to cycle of violence, 110
   operations to interdict, 71
   relative importance, 112–113
   threat to elections, 70–71
Sunni extremists, 93, 95, 104, 113, 140
Sunni Iraqis
   December 2005 election participation, 85–86
   disenfranchised sentiment among, 82
   engagement with, 68, 85–86
   initial threat assessment in 2004, 27
   January 2005 election nonparticipation, 43–44
   rejectionists, 24, 27
   voting against new constitution, 74
surge
   approval of, 145–146
   rollout, 146–147
   views on additional forces, 142–144

Tal Afar–Mosul corridor operations, 70
Task Force 134, 36
Task Force 6-26, 36
threat assessment, initial, 13–14, 25, 27, 37–38, 46–47
training
   of indigenous forces, 164
   train the trainers, 61
transition concept, 56–61
   condition-based, 76
transition readiness assessment (TRA), 72
transition teams
   concept, 58, 60
   early results, 57–58
   effectiveness, 71–72
transitions
   adaptation during, 55–56
   disruption from, 55
   maintaining momentum, 172–174
   momentum loss, 54, 67–68
troop levels
   2004 mid-year, U.S. and coalition, 8, 21
   2006 drawdown cancellation, 113–115
   2006 drawdown plans, 83, 107
   additional for 2004 election, 43
   for embedded advisors, 58
   surge in, 142–143, 145–147
trust, establishing with civil leaders, 123

understand, visualize, decide, direct process, 154–155
United Nations Security Council Resolution 1546, 6, 9–10, 29, 36, 48, 49, 53, 56–57, 62, 66, 75, 165, 192, 197
   consultative mechanisms, 16
   mission description, 157
   timetable, 9–10

INDEX

United States Central Command (USCENTCOM) commander, 7, 19–20, 157, 165–166. *See also* Abizaid, John, GEN USA
Unity, Security, and Prosperity plan, 100
unity of effort
  importance of, 156–159
  One Team/One Mission concept, 10–11
U.S. Army Corps of Engineers, Gulf Regional Division, 36
U.S. Army partnership alignment with Iraqi army, 59
U.S. Army War College, 8

Vines, John R., LTG USA, 3, 63, 64
visualizing, process of, 155
voter participation, 43–44, 74, 85–86

Washington consultations
  December 2004, 48–49, 53
  January 2005, 64–66
  December 2005, 81–83
  June 2006, 104–108
  October 2006, 123–124
  December 2006, 145–146
Western Euphrates River Valley campaign, 70–71
"Wonder Bread" chart, 25, 27

www.ingramcontent.com/pod-product-compliance
Lightning Source LLC
Chambersburg PA
CBHW080539170426
43195CB00016B/2613